BRITISH DE

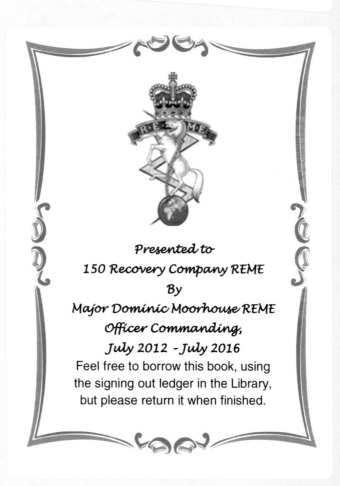

Presented to
150 Recovery Company REME
By
Major Dominic Moorhouse REME
Officer Commanding,
July 2012 - July 2016
Feel free to borrow this book, using
the signing out ledger in the Library,
but please return it when finished.

British Defence Policy

Striking the Right Balance

John Baylis
Senior Lecturer in International Politics
University College of Wales, Aberystwyth

MACMILLAN

First published 1989

Published by
THE MACMILLAN PRESS LTD
Houndmills, Basingstoke, Hampshire RG21 2XS
and London
Companies and representatives
throughout the world

Typeset by Latimer Trend & Company Ltd, Plymouth

Printed in Great Britain by
Camelot Press Ltd, Southampton

British Library Cataloguing in Publication Data
Baylis, John, *1946–*
British defence policy: striking the right
balance
1. Great Britain. Defence. Policies of
government
I. Title
355'.0335'41
ISBN 0 333 49132 7 (hardcover)
ISBN 0 333 49133 5 (paperback)

To Marion, for her 'generosity of spirit'

Contents

Acknowledgements

British officialdom is notoriously secretive, especially on matters of security. Unlike academics across the Atlantic, there is little opportunity in Britain for 'outsiders' to involve themselves in the policy-making process. One of the few ways of trying to get the necessary 'feel' for the subject is through personal contacts. Over the past three years while I have been working on this book, I have had the privilege of interviewing, and corresponding with, a wide range of individuals who are currently involved in, or have in the past been involved in, the formulation or execution of British defence policy. These contacts have proved to be invaluable. Numerous individuals have been kind enough to share their experiences with me and provided me with insights which I could not have acquired from any other source. Almost without exception, however, the officials and officers concerned would only talk to me on the condition that they remained anonymous. To all these individuals I am very grateful indeed.

I wish also to express my thanks to the following people: to my colleagues Professor John Garnett, Professor Ken Booth, Dr Colin McInnes and Dr Moorhead Wright for all their help with various aspects of the book; to Lord Carver who has patiently replied to all my enquiries; to the Rt Hon. Leon Brittan for permission to quote from his study, *Defence and Arms Control in a Changing Era*; to Mr John Day of the Policy Studies Secretariat for setting up a series of interviews in the Ministry of Defence; to the library staffs of the Hugh Owen Library, Chatham House and the International Institute for Strategic Studies; to the staff at the Public Records Office at Kew; to Simon Winder, my editor, for his faith in the author and his professional approach during the publication of the book; to Keith Povey for his editorial skills in the production of the book; to Mrs Shelagh Byers for her thorough and painstaking compilation of the statistics for the appendices; to Alan Macmillan for cheerfully undertaking the task of producing the index; and to Mrs Susan Davies who typed a large proportion of the book and who helped to introduce me to the complexities of the word processor!

Finally, and most importantly, I am grateful to my family for their unfailing encouragement and support. In particular, this book is lovingly dedicated to Marion, who continues to perform the myriad tasks of wife, mother and teacher with great fortitude, patience and good humour.   JOHN BAYLIS

Preface

Most of the literature on British defence policy since 1945 is highly critical. Titles such as *The Long Retreat, The Long Recessional, The Collapse of British Power* and *A Nation in Retreat?* reveal a widely held view that the United Kingdom's post-war security policy has been characterised by a continuous process of contraction and decline.[1] Readers are left with the strong impression that British defence policy has been largely unsuccessful. The critics are divided about who is responsible for this failure. For some the British economy is the main 'villain of the peace'.[2] For others the blame lies squarely with the governments of both parties. They are accused of such things as 'failing to clarify priorities', 'avoiding choice', 'ducking the problem', failing to engage in 'serious rethinking', as well as suffering from 'an absence of political vision', and 'a lack of clarity and consistency of purpose'.[3] Above all the problem, we are told, lies with the governments' incrementalist and ad hoc approach to defence policy-making. The long retreat from power has been conducted 'in the British tradition of pragmatism and empiricism with little regard to long term planning'.[4] This negative interpretation of British defence policy-making is summed up well in Martin Edmonds critique, in which he says:

> The essential problem that Britain has faced throughout the post-war period is that there has been insufficient resources to meet the country's defence commitments as defined by successive governments. In trying to meet these commitments through expedient compromises, time slippages and piecemeal economies, successive governments have contributed to the over-riding problem that the balance of policy and material, men and organisation has never been properly achieved.[5]

The implication of these criticisms seems to be that if only more attention had been given to long-term planning and a systematic process of establishing priorities, British defence policy would have been better, more 'appropriate', than it has been. But would it? Is 'incrementalism' or 'muddling through' such a bad thing? Has British defence policy in fact been unsuccessful? Does the establishment of clear and unambiguous priorities always produce a better defence policy? In the present debate about future defence policy, the need for a radical review of defence policy and clear emphasis on priorities is taken for granted by many academic defence analysts and opposition politicians. The Conservative government, however, holds that such a major review is unnecessary. Pragmatism will see us through. The Ministry of Defence, we are told, will get by with 'adjustments to the programme' and 'better value for money'.[6] Who is right? Which approach is most likely to bring about the best defence policy for Britain in the future? Are these approaches necessarily mutually exclusive?

These are the questions which this book attempts to explore. The initial task is to try to explain the reasons why incrementalism has tended to characterise Britain's approach to defence. This involves an analysis of the particular problems associated with defence, the nature of British strategic culture, and the British approach to policy-making in general and defence in particular. The main section of the book considers how incrementalism has affected British defence policy over the post-war period. Attention is focused on how British policy-makers have attempted to resolve a number of recurring questions and dilemmas by striking what they regard as the right balance. Chapters 1 to 5 deal with these recurring questions and dilemmas.

1. Should Britain's main contribution to European security be on land (the continental commitment) or at sea (the maritime strategy)?
2. Should the nation concentrate on threats to European security or be prepared to intervene in conflicts elsewhere in the world?
3. Should the 'special relationship' with the United States take precedence over defence cooperation with Britain's European allies?
4. Should priority be given to strategic nuclear forces, if necessary at the expense of conventional forces?
5. Should the nation supplement its power by playing a part in an integrated military alliance or retain its national independence?

Chapters 6, 7 and 8 then attempt to make an assessment of the contemporary and future challenges to British defence policy. In the light of these economic, technological and political challenges, the final two chapters consider the strengths and weaknesses of Britain's 'traditional' approach to defence decision-making and suggest how some improvements might be made.

Introduction: The Incrementalist Approach to Defence Policy

The incrementalist approach to decision-making is usually associated with the work of two social scientists, David Braybrooke and C. E. Lindblom. According to Braybrooke and Lindblom the purpose of governing should be to keep the options open by piecemeal choice. In their book, *A Strategy of Decision*, they take issue with the cost–benefit analysis approach to decision-making which derives from economics.[1] In this approach all the costs and benefits which arise from a programme (both monetary and non-monetary) are identified in order that policy-makers can compare benefits to cost and so decide whether a policy is worthwhile, or which among several competing alternatives is most worthwhile. Braybrooke and Lindblom argue that such an approach undermines flexibility in decision-making by closing off options which might prove to be useful in the longer term. Richard Rose has described the distinction between the two approaches in this way:

> Instead of attempting to spell out in detail all the presumed causes and effects of alteratives in order to determine a complete and coherent best-value program, governors are advised to move one small step at a time, avoiding long-term choice in favour of a sequence of short-term choices that allow them to reverse field when this appears immediately desirable.[2]

The name given to this approach by Braybrooke and Lindbolm is 'serial disjointed incrementalism' (more popularly known as 'incrementalism' or 'muddling through').

The term 'serial disjointed incrementalism' derives from a number of assumptions inherent in this approach to decision-making. The first assumption is that political choices are usually *incremental*, 'involving limited, small-scale changes'. In steering the ship of state, governments usually prefer to nudge the tiller this way and that rather than exercise dramatic changes of direction. A second assumption is that decision-making usually takes on a *serial* form, in the sense that most decisions invariably involve an almost never ending sequence of action. 'If harried decision makers can spread a large number of separate decisions over time, they need not decide a program's ultimate objective.'[3] Even when long-term goals are clear decision-makers may not wish to follow a coherent and detailed strategy. 'Taking one step at a time' towards a destination is often a preferred approach. A third assumption is that a series of choices made independently often appear *disjointed,* 'some going sideways or backwards from the direction implied by preceding choices'. According to Braybrooke and Lindblom this disjointed nature of decision-making is useful because 'it permits the prompt correction of mistaken choices'. Richard Rose makes the same point when he argues that 'changing direction is clearly preferable to consistency in pursuing a policy that is heading for disaster'.[4]

The notion of keeping options open is at the heart of the incrementalist approach. Piecemeal choice is seen as preferable to the 'once-and-for-all decision' which tends to characterise the cost-effectiveness or 'best-value' approach. Because policy-makers cannot know all the causes and consequences involved in a prospective programme, there is nothing to be gained, and a great deal to be lost, so it is argued, by pursuing the most cost-effective choices as if the future could be known with any precision. For most governments a series of small successes or failures is much better than the risk of one really big mistake. The incrementalist approach hedges against the disaster of a grand mistake. This can be illustrated by American involvement in the Vietnam War. Under Eisenhower and Kennedy American military advisers were operating in Vietnam but the United States was not irrevocably committed to the war. President Johnson, on the other hand, made the crucial decision to escalate the war after the Bay of Tonkin incident. The consequences of this decisive step were to be very harmful to Johnson personally and to the United States.

Whether the cost-effective approach or the incrementalist approach (or some combination of the two) is the best for defence decision-making is the key question raised by this study. Although cost-effectiveness has been a technique used in British defence policy, especially in the procurement process since the mid sixties, the overall approach has been essentially incrementalist. With the exception of the withdrawal from east of Suez the main pillars of British defence policy – the nuclear deterrent, membership of NATO, the close relationship with the United States, and the provision of independent military capabilities for operations outside Europe – have remained constant throughout the post-war period. Sweeping changes have been avoided and piecemeal adjustments have been favoured. Before we can consider how this incrementalist approach has affected British defence policy we need to consider why 'muddling through' appears to have been so pervasive.

THE PROBLEM OF DEFENCE

The defence of the realm is the primary responsibility of all governments. On the surface, the task of the policy-maker is straightforward. He has to decide how much to spend on defence and how to allocate the resources available between the different defence roles. As W. R. Schilling has written:

> it is a question of knowing the dimensions of the security problem at hand, the relative importance of the national ends involved, the nature of the means available, and the consequences that would follow for these ends from the alternative ways in which the available means can be employed to secure them.[5]

In practice, of course, such decisions are much harder than they appear. The difficulties of achieving that intangible commodity 'security', the limitations on the resources available, the problems of threat assessment, the impact of technological change, and the uncertainties of the future, all make the task of the policy-maker in the defence field highly complex.

The attempt to match resources to the achievement of security raises immediate difficulties. Security is very much a subjective phenomenon. One person's security is another's insecurity. Given that there can never be a feeling of absolute security (at least in the kind of international system in which we live) and that there are always limitations on the resources available for defence (even for the superpowers), what for one policy-maker will be an acceptable level of security may for another be totally unacceptable. Such questions are after all matters of value judgement and as such there can be no one definitive answer to the problem of how much to spend on defence. Again, as Schilling has argued, the inevitability of equally plausible but conflicting factual premises insures that there cannot be 'determinate answers to questions regarding the consequences for this or that value if one or another course of action is followed'.[6]

Part of the problem of achieving what might be described as 'adequate security' arises from the difficulties associated with threat perception. There is the perennial problem of whether to focus on intentions or capabilities in attempting to identify potential adversaries. Both are difficult to evaluate. How can intentions, which can quickly change, be reliably interpreted? Can capabilities be divided into offensive and defensive categories? Should worst-case analyses or most-probable-case analyses be employed? The opportunities for widely different interpretations clearly exist.

Linked closely to the problem of threat assessment is that of technological change, both for one's own state and for potential adversaries. Given different rates of technological development, fears always exist that rival states will suddenly achieve a breakthrough or a decisive technological edge which will undermine one's own security. There is also the difficulty of long lead-times to deal with. Decisions have to be taken today on weapon systems which may take up to twenty years to produce when the technological environment will almost certainly be very different.

The impact of technolgical change is just one aspect of the wider problem of uncertainty which faces policy-makers. By definition the future is unknowable. It is impossible 'to predict with assurance which of the nation's purposes will be challenged or how or when. Nor is there any ready calculus through which probabilities of occurrence can be safely related to quantities of provision'.[7] It is difficult to know if alliances will last or common interests binding states will endure. Domestic and international situations are not subject to accurate forecasting over short periods, let alone over longer periods.

Above all the problem of defence is the problem of uncertainty. Faced with the uncertainties about security, of how much, or often how little, to spend on defence, of how to assess threats, of how to incorporate technology which is frequently changing into defence plans and how to adapt to political and economic changes at home and abroad, it is not surprising that military planners should try to keep their options open by providing a range of capabilities to deal with problematical contingencies. This is what Samuel Huntington refers to as 'strategic pluralism'.[8] In this sense the problem of defence in general contributes in an important way to an incrementalist approach to questions of security.

BRITISH STRATEGIC CULTURE

There is, however, more to British incrementalism than the general problem of the uncertainties of defence planning. There would also seem to be something in the specifically *British* approach to defence and security which contributes to this style of defence decision-making. The term 'strategic culture' usually refers to the distinctive modes of thought and action with respect to the role of force as an instrument of foreign policy, which derives from geographical, historical, economic and other unique influences.[9] According to Colin S. Gray, there is a tendency for policy-makers in particular states 'to make policy in ways and substance that are congruent with the parameters of their own particular culture'.[10] If this is so, what is the nature of British strategic culture?

A number of students of British history have attempted to identify underlying general principles that have traditionally governed the conduct of British foreign and defence policy. Leon Epstein has argued, for example, that 'historically, Britain has had two major international concerns. The first has been to maintain ocean access to the Empire, the second to prevent any one power from dominating the continent of Europe.'[11] In the literature on British strategy it is these twin concerns with sea power and balance of power politics which figure most prominently.

The traditional importance of sea power in British foreign and defence policy was remarked on by Sir Eyre Crowe in his famous 1907 memorandum. In his view:

> The general character of England's foreign policy is determined by the immutable conditions of her geographical situation on the ocean flank of Europe as an island state with vast overseas colonies and dependencies, whose existence and survival as an independent community are inseparably bound up with the possession of preponderant seapower.[12]

In his memorandum Crowe was echoing a sentiment which went back at least to the Elizabethan period. Francis Bacon's claim that 'He that commands the sea is at great liberty and may take as much or as little of the war as he will' had long been a principle accepted by British policy-makers.[13] Equally influential was Walter Raleigh's notion that 'Whosoever commands the sea commands trade, whosoever commands the trade of the world, commands the riches of the world, and consequently the world itself'.[14]

Sir Eyre Crowe also argued in his 1907 memorandum that England had 'a direct and positive interest in the maintenance of the independence of nations, and therefore must be the natural enemy of any country threatening the independence of others, and the natural protector of the weaker communities'.[15] Although, as Morehead Wright has written, 'the tradition of England as holding the balance has taken on an almost mythical proportion', it is a myth which has had a powerful and continuing influence on the conduct of British strategy.[16] F. S. Northedge has stated that this traditional British interest in balance of power politics derives from two sources. Firstly, the persistent realisation in Britain that any state dominating the European Continent could afford to disarm on land and concentrate its resources on sea (and later in air) power, thus jeopardising British security. And secondly, the assumption that 'repose in Europe resulting from equilibrating the European powers one against the other was essential if Britain was to have her hands free to defend overseas territories'.[17]

It is often argued that from these twin concerns, namely sea power and balance of power politics, other traditional principles of British foreign and defence policy have arisen. These include such things as an interest in 'stability and the status quo', an 'indifference to the internal affairs of other states' (unless they threatened the balance of power), 'a concern to harmonize the interests of as many states as possible', a tendency to think in 'power terms' rather than moral or ideological terms, and an aversion to long-term alliances or permanent commitments.[18] These traditional concerns do tell us something about British strategic culture. The question of whether such generalised principles provide a wholly accurate guide to the conduct of British foreign and defence policy, however, is a matter of some dispute.

If the question of a supposed traditional British aversion to alliances is taken as an example, the difficulty of establishing such general principles can be seen. There is no doubt that the literature on the history of British foreign and defence policy contains numerous references to Britain's 'reluctance to enter binding commitments'; her 'traditional line of non-commitment'; her 'traditional anxiety to avoid commitments'; her 'time-honoured refusal to enter binding alliances'; her 'tradition of eschewing alliances and avoiding guarantees'; her 'tradition of a free hand'; and her 'time-honoured principle of

isolation'.[19] Nor is such a view confined to books on British policy.
Foreign Secretaries throughout the nineteenth century expressed their
unwillingness to give any pledge that would bind Britain to go to war
should a given contingency arise in the future. As Joseph Chamberlain told Hatzfeldt, the German ambassador, in 1898 'the policy of
this country for many years has been a policy of isolation – or at least
non-entanglement in alliances'.[20]

Interesting as these contemporary expressions of opinion are, they
constitute no sort of proof that Britain really enjoyed such a freedom
of action in her foreign and defence policies.[21] It is true that for a
considerable period successive British governments did refrain from
involving the country in long-term commitments which would have
bound the nation to go to war in Europe. At the same time, it would
be wrong to argue that Britain made no commitments during the
nineteenth century. Christopher Howard has argued that, wisely or
unwisely, successive British governments entered into prospective
engagements with a number of powers, both great and small, that
unequivocally entailed the *casus belli*.[22] Despite the wariness of long-
term commitments, it is clear that the traditional principle of non-
involvement is not as absolute as many writers and officials suggest.

The danger of generalising from a selective analysis of history is
also shown from the works of a number of eminent British strategists.
In 1932, Sir Basil Liddell Hart wrote his famous study, *The British
Way in War*, in which he attempted to show that there was a
distinctively British practice of war based on 'experience and proved
by centuries of success'.[23] In his view this approach to war was based
on 'mobility and surprise' and was quite unlike continental theories of
war founded on land power. The central importance of sea power,
Liddell Hart argued, had allowed Britain to conduct a successful
'indirect strategy' against European powers without becoming
embroiled in major land conflicts on the Continent.[24] Cromwell used
an 'indirect strategy' against Spain by raiding her treasure fleets. In
the wars against Louis XIV British sea power was used to cripple the
French economy. In the War of the Austrian Succession sea power is
said to have nullified the French victories on land. We are told that in
the Seven Years War Chatham 'vigorously carried out a grand
strategy that became the purest example of our traditional form'. In
this instance direct military efforts on land were replaced by 'subsidies
to allies and undercover of direct sea pressure on France, indirect
military action was applied to the overseas roots of French power'.
Similarly, Liddell Hart argues that in the Napoleonic Wars 'we

eschewed the main theatre of war and employed our land-based forces for sea-based operations against the enemy's vulnerable extremities'. As a result Britain was able to bring the conflict to an end 'without a British Army setting foot in the main theatre of war'.[25]

In his espousal of the maritime school of strategy Liddell Hart was echoing some of the ideas of the naval historian Sir Julian Corbett.[26] There are, however, some differences between the ideas of Corbett and Liddell Hart. As a result of his study of maritime war from the Elizabethan to the Napoleonic Age, Corbett became convinced that a maritime strategy was not an alternative to a continental strategy (as Liddell Hart was later to suggest). Rather it was an extension of it. Maritime power alone, Corbett believed, was not enough. Military power was also important. The task of British strategists was to harmonise naval and military power. Corbett went on from this, however, to argue that maritime strategy made possible '*the application of limited power even to achieve the unlimited objective of total victory*'.[27] In other words, control of the sea enabled Britain to select a theatre of operations where a limited effort could effectively assist the larger operations of continental allies without being subordinate to them. Whereas for Liddell Hart the British way of war was 'mobility and surprise', for Corbett it was 'the application of limited force to the attainment of unlimited objectives'.[28]

Despite the influential nature of the writings of Corbett and Liddell Hart, both their assessments of the 'British way of war' can be challenged. The traditional tendency for Britain to adopt a peripheral strategy was not the result of choice but of necessity.[29] In the case of the Seven Years War England did not have an ally directly across the Channel or the North Sea, and, in the final analysis, victory was less the result of Pitt's strategy than luck. In the Napoleonic Wars, on the other hand, Britain did fight in the Low Countries whenever they had allies to fight for. It was only after Napoleon had established control of the area (after the war of the Second Coalition) that Pitt the Younger adopted a strategy based on raids, diversions and attacks on secondary fronts. Wellington fought in the Peninsula just as the Eighth Army fought in the Western Desert because it was the only place it could fight. As soon as opportunities arose in the Low Countries in 1815 British forces were sent without hesitation.[30]

Michael Howard has provided an interesting assessment of the strategic analysis of Liddell Hart and Corbett. In his view if there was indeed such a thing as 'a British way in warfare', it was the outcome of a continuous *dialectic* between the maritime school and their 'continental opponents'.[31] Britain's geographical position as an island separated from, while at the same time being part of, the European land mass meant that this dialectic was central 'not only to her strategy but to her political economy and indeed her culture throughout her historical experience'.[32]

Howard's own assessment of the British experience of war would appear to provide a much more accurate insight into the nature of British strategic culture. In his opinion two main conclusions can be drawn from the British experience. Firstly, 'a commitment of support to a continental ally in the nearest available theatre, on the largest scale that contemporary resources could afford, so far from being alien to traditional British strategy was absolutely central to it'. He accepts that the flexibility associated with sea power did create the opportunities for colonial conquest, trade wars, help to allies in Central Europe and minor amphibious operations. These, however, in his judgement 'were ancillary to the great decisions, and they continued to be so throughout the two world wars'.[33]

His second conclusion is that when Britain did pursue a purely maritime strategy, it was always as a result, not of free choice or atavistic wisdom, but of *force majeure*. It was a strategy of survival rather than victory. In this sense sea power can be considered as something which gave Britain a breathing space. It allowed 'us to run away' and escape 'the shipwrecks which overtook our less fortunately-placed Continental neighbours' while we searched for new allies. This method of 'taking as much or as little of the war as one will', recommended by Corbett, is not to be despised, Howard says, but it never, in itself, enabled Britain to win.[34]

What the more subtle analyses by Michael Howard and Christopher Howard suggest is that great caution is necessary in trying to identify simple traditional principles of British foreign and defence policy. While certain traditional themes can be identified in the British approach to strategy, necessity has often been more important than choice. Pragmatism would appear to have been a more powerful element in the practice of British strategy than the pursuit of grand theoretical principles. This was very much the view of the 15th Earl of Derby when he observed on 31 May 1885: 'I do not believe that it is possible for us to lay down any formula or any general rule which shall bind us in our foreign policy for all time and on all occasions. We must deal with the circumstances of each case as it arises.'[35]

This emphasis on pragmatism is one of the most distinctive elements in traditional British strategic thought. In the early 1960s, an

American author, De Witt Armstrong, expressed the view that 'the British strategist ... treats concrete situations within a context; he is most reluctant to generalize about the same situations in the abstract'.[36] In strategy as in law, Armstrong argues, Britain prefers the particular to the general, the precedent that is known to work to the encompassing general rule. Much the same point has been made by John Garnett in the contrast he has drawn between American and British strategic thought. In his view there is 'a fundamental, almost timeless quality about American strategic theory, while its British counterpart is very firmly related to specific problems facing the British government'.[37]

If this assessment of the nature of British strategy is correct, both in terms of substance and style, it helps to tell us quite a lot about the conduct of British defence policy since 1945. In particular, the incrementalism of British policy can be seen to have very strong roots in the pragmatism which appears to be inherent in the traditional approach to strategy.

THE BRITISH POLICY STYLE

Another explanation for the incrementalist approach to defence policy may be found in what a number of writers have suggested is a distinctive British approach to policy-making in general. Political scientists have long been interested in attempting to identify different national policy styles. The term 'policy style' refers to 'the main characteristics of the ways in which a given society formulates and implements its public policies'.[38] Although states do not always formulate policies in exactly the same way, it does seem possible to identify a predominant style. In the case of Britain, studies by Hayward, Jordan and Richardson all seem to agree that there are certain distinctive characteristics about the British approach to policy-making.[39] These include 'a predilection for consultation, avoidance of radical policy change and a strong desire to avoid actions which might challenge well-entrenched interests'.[40]

In their study of British policy style Jordan and Richardson identify what they describe as 'bureaucratic accommodation'.[41] This involves a system of decision-making between groups and government departments in which the mode is one of bargaining rather than imposition. They use the term 'the logic of negotiation' to describe this style. Underlying this consultative aspect of the British approach, they argue, is 'a broad cultural norm that the governing should govern by consent'. Lord Rothschild makes much the same point when he argues that in Britain there is a 'beatification' accorded to compromise.[42]

The policy style of consultation and negotiation inevitably has an impact on the nature of policy which results. It tends to lead to incremental policy change, irrespective of the party forming the government at any one time. This tends to lead to consistency rather than radical policy change. According to Richardson and Jordan, the British system is characterised by 'group constellations' which surround particular policies, acting like a magnetic field, holding the policy in place. In their view:

> It is possible under certain circumstances, to induce policy change (to change the 'magnetic field') but there is a natural predisposition for the field to hold the policy in place. Moreover departments have little desire or incentive to disturb a magnetic field once it has been established. The bureaucratic preoccupation tends to the minimization of disturbance, the securing of a stable environment or negotiated order, rather than a significant policy change.[43]

Even when change is necessary the tendency is to retain the major thrust of policy as far as is possible.

An important question for this study is whether the predominant style of British decision-making is as characteristic of defence as it is of other areas of policy. Because of the secrecy involved, defence is not characterised to the same extent as other areas of governmental policy-making by wide-ranging group-departmental consultation and negotiation. Defence tends to be characterised by a process of 'internalised' policy-making. Policy tends to emerge without much large-scale organised outside group deliberation. The Ministry of Defence nevertheless does have large numbers of advisory committees. The military–industrial complex does have access to the Ministry. And there is a strategic studies community in the universities and professional research institutes which make an input into the very private deliberations which take place on defence policy. Even though the scale of this insider access may not be as great as in many other areas of government, policy which is evolved internally, in the longer term, is only likely to be tenable if it can be 'sold' to influential constituencies both inside and outside the government apparatus. The fact that there has been a great degree of bipartisanship over defence in Britain for most of the post-war period suggests that the processes

of 'bureaucratic accommodation', 'consultation and negotiation' and
compromise are as important in the defence field as elsewhere. Before
this conclusion can be established, however, more attention needs to
be given to the institutions responsible for defence decision-making.

THE DEFENCE DECISION-MAKING PROCESS IN BRITAIN

In Britain, the process of defence policy formulation and detailed
planning has been left to the relevant Cabinet Committee, government departments, and Armed Service Chiefs and their respective
staffs. In reality, these institutions work out the substance, details and
direction of Britain's defence. It is this organisation that is ultimately
responsible for meeting the challenge of balancing the defence equation, of getting policy, capability and organisation to coincide. Any
imbalance in that equation is entirely its responsibility.[44] One of the
major reasons why imbalances have occurred is that the central
organisation of defence has been less a centralised decision-making
body and more 'a confederation of separate power centres'. Constant
post-war attempts to create a more unified and centralised framework
for defence decision-making have run up against resistance, especially
from the Services and Service Departments which have attempted to
retain their autonomy. The record of institutional change which has
taken place since 1945 reflects a continuous attempt to strike the right
balance within the decision-making apparatus itself.

   In the immediate post-war period, the post of Minister of Defence
was created and a Ministry of Defence was set up with executive
authority in certain areas over the three Service Departments and the
Ministry of Supply. Drawing on the experience of the Second World
War (when Churchill had been Prime Minster and Minister of
Defence), the Prime Minister was to act as Chairman of the Defence
Committee of the Cabinet. At the same time, the Chiefs of Staff
Committee and a joint staff system were to be responsible for
operational planning and coordination within the Services. Each
Service Department also was to decide on its own procurement
priorities. As such the 1946 White Paper setting up the new organisa-
tion of defence represented something of a compromise between
Service expertise and the need for unified planning.[45]

   In practice, as General Montgomery predicted, the 1946 organisa-
tional structure proved to be a failure.[46] Without a strong Minister
(and A. V. Alexander was not one) Service disunity and rivalry could

not be kept in check by the new arrangements. The Minister 'could not only be by-passed by the Chiefs of Staff (COS) but also ignored by the three Service Ministers with their independent Parliamentary appropriations'.[47] As a result, during the period of the late 1940s and early 1950s the decision-making process was characterised by intense inter-Service rivalry over defence roles and scarce resources. Defence policy was the result of compromise between the Chiefs themselves and the COS and the government of the day.

The most serious aspect of the institutional framework established in 1946 was that it set a precedent which has left an indelible mark on the central organisation of defence ever since. It established the basic structure with which subsequent reformers have had to work. The fact that the process of reshaping and changing the framework has gone on continuously ever since confirms the widely shared view that it did not provide 'a basis that has served the challenge posed by Britain's defence very well'.[48]

By the early 1950s it was clear that the central organisation of defence established in 1946 was in need of reform. In particular, the rearmament programme undertaken during the Korean War and the Radical Review of defence undertaken during 1953–4 demonstrated the weakness of centralised control. Each of the Services bargained directly with the Treasury for the resources available, with the Ministry of Defence playing only an indirect role. Consequently, inter-Service rivalry played an important role in the decision-making process.[49] While there was general agreement between the Chiefs of Staff and the Ministry on the main outlines of policy – in terms of the commitment to NATO, the nuclear deterrent, home defence and overseas security – the implementation of policy during the mid-1950s nevertheless left a lot to be desired. The process of bargaining and compromise continued to predominate.

In 1955, the Eden government established a new post, Chairman of the Chiefs of Staff Committee, in an attempt to create a single professional voice reporting on defence issues to the Cabinet. When Harold Macmillan took over from Eden after the Suez crisis, he was determined to extend the reforms even further. His experience as Minister of Defence in 1954–5 had convinced him that changes were urgently needed. Duncan Sandys was therefore appointed with a remit to establish much greater central control over the decision-making process. Sandys set about his task in a determined and abrasive manner which resulted in a difficult period of confrontation between the Services and the Minister from 1957 to 1959. The friction created was clearly not conducive to reform. Although Sandys managed to enhance the role of the Chairman of the Chiefs of Staff Committee (now known as the Chief of the Defence Staff), his 1958 White Paper on the *Central Organisation for Defence* did not get rid of the separate Service Ministries as he had hoped to do.[50] As a result of all the controversy Sandys managed to generate, Macmillan decided that the time was not right to push for more radical changes.

Macmillan did, however, appoint Earl Mountbatten to the post of the Chief of the Defence Staff in May 1958. Mountbatten had been a leading advocate of the need for greater centralised control over defence policy for some time, and during his period in office he worked unceasingly to enhance the role of the Ministry of Defence and to downgrade the power of the Service Chiefs and the separate Service Departments. Mountbatten's recommendations (for a single Secretary of State for Defence, two functional Ministers – for Personnel and Research, junior Ministers for each of the three Services and a larger staff for the CDS) predictably were opposed by the other Chiefs of Staff. From their point of view this attempt to create a single Ministry of Defence and abolish the single Service Departments 'would separate power and responsibility to the detriment of accountability'.[51] As professional heads of their Services, with the task of advising the Minister, they believed they ought to retain responsibility for military operations.

Faced with these contrasting views the Defence Minister, Peter Thorneycroft, called upon General Ismay and General Jacob to adjudicate. The result was essentially a compromise in which the Service ministries were incorporated into a single Ministry of Defence, but the role of their CDS was not enhanced and the Service Chiefs retained their responsibility for military operations.[52]

Despite this compromise announced in 1963 the changes which resulted represented an important step forward towards greater centralisation. The Secretary of State was now to assume responsibility for 'all questions of policy and administration which concerned the fighting Services as the instruments of an integrated strategy'.[53] The Secretary of State was to be directly accountable to the Cabinet Defence and Overseas Policy Committee. In this new structure the COS were charged with providing military advice but the principal advisers to the Secretary were to be the Chief of the Defence Staff, the Permanent Under Secretary at the MOD and the Chief Scientific Adviser. At the same time, four new defence departments were established (for operations, requirements, intelligence and signals) and controlled by the COS 'to find the best defence solutions to the problems with which they were faced'.[54]

The struggle for and against further centralisation continued throughout the period of the Labour government from 1964 to 1970. In 1967, pressures on the defence budget brought about another attempt to improve functional rationalisation when the Service ministers were replaced by Under Secretaries of State and two new functional ministerial posts were established, one for Equipment and one for Administration. The Secretary of State, Denis Healey, decided to take the process a stage further in 1968 by replacing the three-Service system of budgeting with a single defence budget system, thereby significantly enhancing centralised financial control. At the same time, a single defence planning staff was set up (replacing the separate Service planning staffs) and the two Ministers for Equipment and Administration established in 1967 were replaced by a single Minister of State for the Armed Forces. A further modification was undertaken in 1971 when, as a result of the Rayner Report, a central procurement executive was established, and the post of Minister of State for Procurement was created.

By the early 1970s the process of continuing adaptation and evolution within the central organisation of defence had produced a structure which represented an uneasy compromise between what Martin Edmonds has described as the 'federalist' and 'functionalist' approaches. According to Edmonds:

> The Ministry of Defence was in effect a federation of three Services with their respective boards and administrative structures on the one hand and, on the other, a grouping of four, strong, functional bodies: for military planning and operations (under the Chief of Defence Staff); science and research (under the Chief Scientific Adviser); central defence administration (under the Permanent Under Secretary); and equipment procurement (under the Chief Executive Procurement).[55]

Up until the early seventies, although the supporters of the federalist approach had managed to restrain the enthusiasm in some quarters for greater functionalism, the trend nevertheless had been very much in that direction. The rest of the 1970s were to witness a slowing of the 'progress' towards greater centralisation. Indeed, for a time power appeared to be drifting back towards the Service Departments. The Heath government between 1970 and 1974 attempted to stem the tide of functionalism by restoring the post of Parliamentary Under Secretary for each of the three Services. As a result, the Services were better able to defend their own separate interests in the battle over resources which characterised the mid-1970s. With Fred Mulley as Secretary of State for Defence under the Labour government from 1974 onwards, the Services also found a champion who tended to favour the federalist over the functionalist approach.

The period of 'constructive erosion', however, was to be short-lived.[56] By the early 1980s the attempt to achieve greater centralisation was resumed. Faced with growing inter-Service rivalry over John Nott's defence review, the Thatcher government decided to get rid of the Service Ministers and replace them with two Ministers, one for the Armed Forces as before, and one for Procurement. In addition, the power of the Chief of the Defence Staff was increased and the influence of the COS as advisers to the Secretary of State and the Defence Council was reduced. The purpose of these reforms was 'to increase the Secretary of State's ability to delegate *functional* responsibilities to his Ministers; to emphasise the defence, as opposed to the single Service, responsibilities of the Department; to strengthen political direction throughout the Department and to allow far greater Ministerial control of the defence procurement process at an earlier stage than in the past'.[57]

The tenure of Michael Heseltine at the Ministry of Defence witnessed an even greater push towards increased functional control. In 1984, Heseltine introduced the Management Information for Ministers System (MINIS) into the MOD which he had used in his previous Department (Environment). The main thrust of the innovation in management was to try to 'provide an over-all picture of all the functions of the Department and to allow them to exercise their management of these functions positively, rather than reactively'.[58] As a result, the emphasis was to be placed even more than in the past on functional rather than Service divisions within the defence decision-making structure. Significantly, during the MINIS exercise the Service Chiefs were kept very much on the side-lines and the results were presented to them as a *fait accompli*. The most important of the changes involved the CDS and the PUS, who were to become the principal advisers to the Secretary of State. The CDS was to advise on military operations and strategy and the PUS on political and financial policy. In addition, the post of Vice-Chief of the Defence Staff was created and four new functional departments for Personnel, Systems, Commitments, and Strategy and Policy (each under a

Deputy CDS and with its own defence staff). Centralised financial control was to be improved with the creation of an office of Management and Budget under the PUS.

The idea behind these changes was that the role of the individual Chiefs of Staff in the policy and planning process was to be significantly curtailed. They were to remain the professional heads of their Service and as such they retained responsibility for morale, fighting effectiveness, discipline and efficiency. Their function now, however, was to be one mainly of management with little influence over the formulation of defence policy (apart from the retention of their symbolic right of access to the Prime Minister).

These 1984 reforms were the subject of a great deal of controversy. Public criticisms were voiced by a number of ex-Chiefs of the Defence Staff who argued that the process of centralisation had gone too far. In a letter to *The Times*, Field Marshal Lord Carver, who during his period as CDS had favoured greater functionalism, argued that it was an error to imagine that policy and management were two different functions. They were, he suggested, 'inextricably intertwined'. He accepted that there had to be a central assessment and direction of the general balance of effort between different components of defence policy but he went on to argue that

> within the global central allocation of priorities and resources, the single Service machinery is best qualified to balance all the different factors affecting the effort to be provided by its Service: to determine what weapons systems and organisation are required, how many units there should be, how they should be organised, equipped, trained and accommodated.[59]

Lord Carver's view was that the permanent balancing act was highly sensitive to a large number of factors, which could only be accurately assessed as a result of 'accumulated experience and contact with the grass roots'.[60] Such experience was only to be found within the single Services. For Lord Carver, the Heseltine reforms of the central organisation of defence were likely to undermine the ability of the Service chiefs to advise the CDS and relegate them to the 'task of routine management of their Services'.[61]

Despite these concerns over the loss of Service autonomy and the declining influence of the Service Chiefs over defence policy-making, the Heseltine reforms were accepted by the government. Their implementation, however, as usual proved to be difficult to achieve in full.[62]

Despite the government's desire to improve the role of the central policy staff in the defence decision-making process, in practice the Service Chiefs continued to play a very influential part. As a result, it is still too early to say, in the late 1980s, whether the federalist or functionalist models hold sway in the Ministry of Defence despite all the changes which have occurred.

Looking back over the period since 1945 what stands out is the continuing process of adjustment and readjustment which has taken place within the central organisation of defence. The Ministry of Defence and the higher echelons of the Services have been in continuous flux at a time when they have had to carry out frequent reappraisals of British defence policy as the domestic and international environments have changed. Plans have had to be revised to conform to new decisions when the defence structure and the chain of command was continuously melting and reforming. As a result, this constant process of institutional adaptation has played its part in the incrementalist approach to defence decision-making which has characterised the post-war period. Striking the right balance between Service autonomy and centralised control within the decision-making framework has been just as difficult as striking the right balance in the major areas of British defence policy. But what has been the result of incrementalism? To demonstrate the impact of this process we now turn to the five main issues which have dominated the debate about British defence policy in the period since 1945.

1 The Continental Commitment Versus a Maritime Strategy

One of the most important issues in British defence policy in the late 1980s is the debate between those supporting a continentalist strategy and those advocating more emphasis on Britain's maritime contribution to the defence of Western Europe and beyond. As the Introduction has attempted to show this is hardly a new controversy. The dialectic between the maritime school and their continental opponents has been a continuous theme in the history of British defence policy over the centuries, including the first part of the twentieth century. As Michael Howard and Brian Bond have shown, the debate was particularly significant during the period leading up to the Second World War.[1] For much of the inter-war period British governments remained extremely reluctant to make any formal commitment to the defence of Western Europe. The priority given to imperial defence meant that the Chiefs of Staff were only prepared to accept a 'limited liability' in Europe. As the situation in Europe deteriorated in the late 1930s, however, the government's dilemma sharpened. In December 1937, the Minister for the Coordination of Defence, Sir Thomas Inskip, presented a list of defence priorities to his Cabinet colleagues in which the continental commitment ranked last.[2] By March 1939 events in Europe forced the Chiefs of Staff to carry out a remarkable volte-face. They now recognised, rather belatedly, that it was difficult to see 'how the security of the United Kingdom could be maintained if France were forced to capitulate'. Britain, they argued, must be prepared to participate in 'the last defence of French territory'.[3]

The recognition that this conversion to a 'continental commitment' came too late has had a profound effect on post-war defence planners. Even before the war ended the future planning staff in the Foreign Office were producing reports emphasising that Britain's security interests would henceforth have to be bound up with the Continent much more directly than in the past. A paper on 'Security in Western Europe and the North Atlantic' produced by the Post-Hostilities Planning Staff, in November 1944, noted that after the war ended

19

Britain would require powerful allies and 'defence in depth' on the Continent in order to balance 'the greatest land power in the world', the Soviet Union.[4] The conclusion of many of these studies was that if Britain was to continue its traditional balance of power role, it would need to take on the responsibility, when the war ended, of setting up a Western European Security Group.[5]

The key question, however, centred on exactly what Britain's contribution should be to the defence of Western Europe. In the last year or two before the war ended a major debate took place between the Prime Minister, Winston Churchill, and the Foreign Secretary, Anthony Eden, over whether Britain should adopt a continental or maritime–air strategy in the post-war period. Churchill accepted the idea of a Western European Group put forward by the Foreign Office but he had very little faith that it would be able to contribute very much to British security in the foreseeable future. He believed that the small states of Western Europe would be 'liabilities rather than assets' if Britain bound herself to them in a scheme of common regional defence. Churchill sent a minute to Eden on 25 November 1944 in which he argued that there was little to be gained from a Western European Group until France had been able to build up her army once again – which would probably take from five to ten years. The Prime Minister was rather dismissive of many of the small states:

> The Belgians are extremely weak, and their behaviour before the war was shocking. The Dutch were entirely selfish and fought only when attacked, and then for a few hours. Denmark is helpless and defenceless, and Norway practically so. That England should undertake to defend these countries, together with any help they may afford, before the French have the Second Army in Europe, seems to me contrary to all wisdom and even common prudence.[6]

Churchill was not convinced that there had been any fundamental change in Britain's strategic position as a result of the war. There was no overwhelming need for a strong continental presence in the post-war world. Technological changes, he accepted, had damaged Britain's island position but a strong air force and sufficient naval power would continue to be a tremendous obstacle to invasion. In his view, keeping a large army on the Continent would be too expensive. It would be much wiser to put the bulk of the money into the air force 'which must be our chief defence with the Navy as an important assistant'. Churchill, therefore, was not convinced that the events of the 1930s necessitated a fundamental change in British defence policy towards a major peacetime continental commitment.

Eden agreed with some of the Prime Minister's views on the danger of taking on wide-scale commitments on the Continent. In a memorandum on 29 November 1944, he nevertheless argued that the time had come to start serious thinking about the problem of a regional defence system.[7] The Foreign Secretary reasoned that it was important that her allies in Western Europe should not get the impression that Britain was not prepared to accept any continental commitments. If they did come to that conclusion they might be likely to seek accommodation and defence arrangements with the Soviet Union. In Eden's view the advent of long-range missiles really had altered Britain's strategic position and it was thus necessary to achieve some form of 'defence in depth' to enhance Britain's future security. A Western European Security Group, he believed, could provide the necessary depth in defensive terms and the manpower to ease Britain's burden in the post-war world. It would not be necessary to maintain a large standing army on the Continent, though a larger commitment than in the past would be desirable.

This important debate was to continue well into the post-war period with the Labour Party in power. The Foreign Office, now under Ernest Bevin, provided a lead in the consolidation of Western European defence with the signing of the Dunkirk Treaty (with France), in March 1947, and especially with the initiative which led to the signing of the Brussels Pact Treaty (with France, Belgium, Holland and Luxembourg), in March 1948.[8] Despite the strong political commitment made in the Brussels Pact to the defence of Britain's Western European allies, the debate about the precise form the military commitment would take was far from resolved in 1948. The British government was still not 'in a position to give assurances about the role they intended to play in operations on the Continent of Europe'.[9] For defence planners the primary area of British security policy remained the Middle East, as it had been ever since 1945. At a Chiefs of Staff meeting on 23 December 1947, after the failure of the London meeting of the Council of Foreign Ministers, General Ismay had warned that the question of a continental strategy which had so far been neglected by them would have to be considered.[10] When the Chiefs of Staff met on 29 January and 2 February 1948 a major row broke out between them over a paper submitted by the CIGS, General Montgomery, advocating a continentalist strategy to back up Bevin's political initiatives towards Western Europe.[11] The Naval and Air

Chiefs, Admiral Cunningham and Air Marshal Tedder, both argued in favour of a more traditional maritime–air strategy based on small highly mobile forces. This was also the view of the Prime Minister, Clement Attlee. At a meeting later in February to discuss Montgomery's paper, the Prime minister joined Cunningham and Tedder in opposing the continentalist strategy. Montgomery, however, found support from the Foreign Secretary and the Minister of Defence, A. V. Alexander.[12] As a result of this split the debate between the two strategies raged through March and April and it was only on 10 May 1948, in a meeting between the Chiefs of Staff and the Minister of Defence, that it was agreed that Britain should plan to fight a campaign alongside its allies in Western Europe. Even then precisely what contribution Britain should make was far from resolved. As Elizabeth Barker has shown: 'until the end of 1949 no serious military planning for the defence of western Europe had been done except by the Western Union ... Even this suffered from British reluctance to say what land forces, if any, would be sent to the continent in case of war.'[13] Indeed, the formal commitment to send reinforcements to the Continent in the event of war was not accepted by the Defence Committee until 23 March 1950.

Britain's slow, and in some ways agonising, conversion to a continental commitment was therefore not agreed by British Ministers until two years after the Brussels Treaty had been signed. It would be another four years before Britain was prepared to accept a long-term commitment to European defence. This came with the Paris Agreements of 1954. With the threatening crisis over the collapse of the European Defence Community proposals, the Foreign Secretary of the day, Anthony Eden, formally committed British troops to the defence of Western Europe for the next forty years. This commitment which was made for largely political reasons to reassure the French and prevent 'an agonizing reappraisal' of American foreign and defence policy, symbolised Britain's final, formal acceptance of a large-scale, peacetime military presence on the Continent and a willingness to fight alongside Western European allies in the event of war.

From 1954 onwards there were numerous fierce procurement battles between the three Services (especially over Britain's aircraft carriers). There were also frequent differences between the Chiefs of Staff over whether to prepare for a long or short war in Europe. In the late 1950s, the decision was made to scale down the British Army of the Rhine from 77 000 to 55 000 troops. Throughout all this, however, the continental commitment itself remained largely unchanged. This is not to argue that absolute priority has always been given to BAOR throughout this period. Overseas operations and responsibilities have at times made BAOR appear to be 'the forgotten army'. Nevertheless, for political as well as for military reasons no British government has been prepared to reverse the commitment made in 1954. Despite constant debates about priorities for most of the past thirty years the dilemma between the continental commitment and the maritime strategy has not been particularly acute. Throughout the period the Navy has continued to play a key role in the defence of the Channel and the Eastern Atlantic. What governments of both political parties have tried to do is to balance Britain's land and sea contributions to Western defence. The recognition that both contributions were of vital importance was spelled out clearly in the 1974–5 defence review. The Labour government of the day, faced with continuing economic difficulties, identified four priority areas of British defence policy. The strategic nuclear deterrent was one, home defence was another, and the remaining two were defence of the Central Front and the Eastern Atlantic. The continental commitment remained of vital importance but so did Britain's naval contribution to the defence of Western Europe.[14]

The 1980s, however, have witnessed a resurgence of the bitter debate which characterised the late 1940s. In 1981, the Defence Secretary, John Nott, discovered that despite the government's commitment to an annual 3 per cent real increase in defence spending the Ministry of Defence could not live within its budget. As a result of his major defence review, John Nott concluded that 'something had to go'.[15] He rejected the incrementalist 'cheese paring' or 'equal misery' approach in favour of a fresh and radical reassessment of each of the four major priority areas of British defence policy. The decision had already been taken to modernise the strategic nuclear deterrent. In the government's view this force remained 'a crucial and unique ingredient of Britain's defences'.[16] It was further decided that the defence of the home base required more spending not less after years of neglect. The choice, therefore, was between the continental and maritime contributions to European defence. Nott's conclusion was that the land and air forces Britain maintained in the Federal Republic of Germany on the Central Front were so important 'to the Alliance's military posture and it's political cohesion that they had to be maintained'.[17] At a time when Britain was having difficulties with it's European allies over the renegotiation of Britain's contribution to the

EEC budget, it was felt that nothing should be done to weaken its contribution to the land defence of Europe. Such a reduction would have great symbolic importance in signalling a declining British commitment to Europe which would weaken Britain's hand in the negotiations. This was also a time when the Alliance was seeking an improvement in its conventional forces in Europe to raise the nuclear threshold. Once again, cuts in BAOR would undermine this strategic objective.

It thus fell to the Navy to carry the main brunt of the cuts. Nott accepted that Britain must continue to make an important contribution to the defence of the Eastern Atlantic but he concluded that this should henceforth be achieved through the provision of maritime air power and submarines. The same job had to be done, it was argued, but by other (cheaper) means. The surface fleet was scheduled to be cut by 25 per cent. The numbers of destroyers and frigates were to be reduced from 59 to 50 (with 8 of the 50 to be mothballed); two carriers were to be sold or retired; two amphibious ships were to be scrapped, as well as nearly all the Navy's auxiliary landing ships.[18]

Faced with what it regarded as a straight choice between the continental commitment and the maritime contribution to European defence, the Thatcher government had reluctantly agreed, against the advice of its Naval Chiefs, to give priority to the defence of the Central Front. The defence of the Eastern Atlantic was not to be abandoned but, despite the government's denial, it was to be significantly weakened as a result of the 1981 Review.

Subsequent events, however, were to show that the government's dilemma was far from over. The outbreak of the Falklands War in 1982 led to the largest British naval operation since 1945. To the naval lobby in particular (which retained a great deal of support in certain sections of the Conservative party), these events confirmed the 'strategic myopia' of the 1981 review.[19] In the aftermath of the war, the debate about defence priorities, which the Nott review had appeared to close in favour of the continental commitment, was once again reopened. In the years since 1982 the government has faced a barrage of advice from the press, politicians and former serving officers on whether to pursue a completely different strategy towards Western European defence. A key question in British defence policy once again has been over how to strike the right balance in the dialectical relationship between the continentalist and maritime contributions to European defence.

Shortly after the Falklands War *The Times* led a campaign on behalf of the Navy, arguing in a number of editorials that Britain had 'too much of the Rhine'. In August 1983, the paper made three concrete proposals:

1. That the German Army should assume responsibility for the operational sector of Allied Command Europe at present entrusted to the Rhine Army.
2. That the British Corps in the Northern Army Group should be held as a tactical reserve and reduced in size accordingly.
3. That Britain should relinquish command of the Northern Army Group.[20]

Then, in May 1984, a leader article claimed that: 'The peacetime establishment of the Army and RAF in Germany have no tactical rationale. The line-up in Central Europe makes military nonsense. It is born of old political formulae which have outlived their relevance.'[21]

This campaign by *The Times* was supported by a number of influential figures with very diverse political views. Former Prime Minister James Callaghan was one who argued that Britain had taken on too great a burden in Europe. In his view the British Army of the Rhine ought to be halved.[22] On the other side of the political spectrum, Enoch Powell urged the government to reshape the philosophy of Britain's armed forces to re-establish 'the primacy of our maritime element'.[23] Also on the right of the political spectrum the Adam Smith Institute (a key supporter of Mrs Thatcher's monetarist policies) called on the government to make substantial reductions in BAOR. In a report produced in 1983 the Institute argued that British defence policy had given too much priority to the defence of the Central Front. The report advocated a restructuring of the armed forces in order to put more emphasis on reinforcement capabilities, air power and surface ships for the Navy. Only then, it was argued, could Britain get away from the 'Maginot line mentality' of static defences in Central Europe.[24]

Not surprisingly, this need for a reorientation of British defence policy was supported by the Naval Chiefs. What was perhaps more surprising was that they were prepared to go public in their criticisms of government policy. In a speech to the Royal United Services Institute in September 1982 Admiral Sir Henry Leach, the Chief of the Naval Staff and First Sea Lord, argued that it was far better for Britain to concentrate on its traditional expertise at sea. Surely, he said, 'the sensible and cost-effective thing for NATO to do is to build on what already exists; for countries such as West Germany with their continental expertise and geography to concentrate on the Central Front – after all it is their own soil – and for the United Kingdom to maintain its lead at sea'.[25] The Admiral even went as far as to suggest that the government's defence policy was nothing more than 'a major con-trick involving a catalogue of half-truths'.[26]

Admiral of the Fleet Lord Hill-Norton, a former Chief of the Defence Staff, was even more outspoken in a chapter he wrote for a book in 1983. In the opening paragraph Lord Hill-Norton argued that:

> The Defence Policy outlined in the White Paper of 1981 and 1982, and to a large extent restated in that of 1983, is based on several fundamental misconceptions, of which the most dangerous (certainly in the long term) may be briefly summarised as: 'Our front line is in Germany where lies the greatest threat to the United Kingdom; war, if it comes in Europe, will be short, so the Atlantic (never mind the distant seas) hardly matters, and where it does it can be defended by a handful of nuclear submarines and long range maritime patrol aircraft'. *Virtually every word of this is demonstrable rubbish*; it flies in the face of history and the NATO assessment of the total threat.[27]

According to Lord Hill-Norton's strategic analysis new weapons and new military techniques required new tactical doctrines in Europe. The battlefield balance was moving more towards effective defence and this would facilitate more economical deployment of forces in Europe. As a result Britain's contribution to the defence of the Central Front would need to be radically restructured and could be reduced by between one-third and one-half. Echoing Sir Thomas Inskip's 1937 list of priorities, Lord Hill-Norton suggested that Britain's contribution to the land defence of Western Europe should 'flow from what can be made available *after* the first priority has been given to the creation of [maritime] forces of the right size and shape'.[28]

In this public debate about priorities (which clearly reflected the more private debate between the Chiefs of Staff themselves), the naval lobby did not have it all its own way. Other eminent former Chiefs of the Defence Staff from the Army and Air Force also aired their views on every occasion possible. Lord Carver and Lord Cameron in particular went out of their way to argue that the defence of the

Central Front had to remain a crucial priority.[29] In their view, the Falklands War had demonstrated, not the utility of surface ships, but their growing vulnerability and the problems which this created for conventional convoy operations in the Eastern Atlantic. According to those supporters of a continentalist strategy it was important to maintain the existing forces and equipment on the ground in Europe. Once hostilities began it would be weeks before the convoys could have any impact on the war and by the time they arrived it might well be too late if the first line of defence in Western Europe was neglected. In supporting the main outline of the 1981 review both Lord Cameron and Lord Carver accepted that 'a degree of maritime warfare capability is essential to the security of Europe'.[30] They also recognised that Britain was best placed of all the European members of NATO, through her history and geographical position, to make a major contribution to NATO's maritime forces. This could best be done through the provision of submarines and long-range maritime aircraft rather than surface ships; and it was crucial that *any* provision of maritime forces 'must not be at the expense of an adequate contribution to a continental strategy'.[31]

It is interesting to note that in this continuing debate over the continentalist and maritime strategies both sides have used history and Britain's traditional experience to justify their arguments. For Lord Hill-Norton, Admiral Sir Henry Leach, Enoch Powell and other supporters of the naval lobby like John Wilkinson and Michael Chichester, Britain should once again give priority to its traditional expertise at sea. Like Corbett and Liddell Hart they emphasise sea power as the traditional British approach to warfare. The continentalist reply is that history has a more important lesson: while sea power may have been important Britain has had a traditional and vital interest in influencing events on the Continent. Lord Carver echoes Michael Howard when he argues that:

> For centuries the basis for the protection of the interests of Britain has been the need to ensure that the continent of Europe is not dominated by a power that is unfriendly to us. That was the basis of opposition to France in the time of Louis XIV and XV and of Napoleon Bonaparte, and to Germany under Kaiser Wilhelm II and Hitler. It is the fundamental reason why we welcomed the foundation of NATO, the threat from Germany having been replaced by that from the Soviet Union. Throughout our history a rival view of the priority for our strategy has been urged: that we

should turn our backs on the continent and concentrate our efforts on securing trading advantage and access to raw materials across the oceans, in the Pacific and Indian Oceans, the South and Western Atlantic. That strategy brought some significant successes, as well as some notable failures; *but it has never been able to preserve our fundamental interests and security, if alliances on the continent of Europe have failed us.*[32]

For both schools of thought, in the debate which has raged during the 1980s, history does not suggest the abandonment of one role or the other. Rather their interpretation of history suggests an emphasis in favour of one role as opposed to the other. This question of relative priority has been at the heart of government's dilemma in recent years. Faced with the legacy of the Falklands War and the public and private disagreements between the Service Chiefs since the early 1980s, the government has struggled to avoid the kind of clear decision made by John Nott in 1981 and instead has attempted to balance the continentalist and maritime lobbies. The Defence White Papers since the Falklands War have attempted to defuse the controversy by arguing that although the main priorities established in 1981 remain in force, the government intends to retain rather more surface ships than envisaged in the Nott review. One of the key questions in the defence debate in recent years has been whether the government has significantly revised its ideas about the Navy or whether it has simply suited its purpose to let that notion gain currency. While the government has continued to wrestle with the dilemma it has repeatedly pledged to maintain a Royal Navy force of *around* fifty frigates and destroyers as well as to replace or refurbish the amphibious assault ships, HMS *Fearless* and HMS *Intrepid*, which were to be phased out under the 1981 review.

In contrast to the 1981 review, the government's approach to the dilemma of the continentalist versus maritime strategies in recent years has been to pursue a traditional incrementalist style of decision-making. Faced with only marginal increases in the defence budget in the future, it seems likely (as Chapter 6 argues) that hard choices are going to have to be made again. One of the key questions facing defence planners at present is whether to do what was done in 1981 and attempt to establish clear priorities significantly cutting the continental commitment or the defence of the Eastern Atlantic (or one of the other two key roles in British defence policy), or whether the incrementalist approach which at present prevails should be continued. This question of how to strike the right balance between the continentalist and martime contributions to European security is a vitally important one for the future of British defence policy. Which decision-making approach is adopted is likely to have a major influence on what balance is struck. We will return again to this question in Chapters 9 and 10.

2 European Versus Global Defence

The debates about the continental commitment and a maritime strategy are closely related to the question of whether priority should be given to European or global defence. Maritime forces designed to help defend the Channel and the Eastern Atlantic can also be used to promote and protect British (and Western) interests outside the European theatre. In an important sense, part of the controversy between land–air and sea–air contributions to European defence has focused on the wider question of intervention capabilities for national and alliance purposes outside the European theatre.

Many of the studies of post-war British foreign and defence policy have focused on what is described as the illusion of Great Power status.[1] According to this popular thesis one of the major problems of Britain's external policy up until the late 1960s (and some would say right up to the present day) has been that she has failed to adjust to her diminished position in the world. The legacy of her imperial and colonial past has produced an image of her role in the world quite out of proportion to her reduced circumstances. Britain, it is often said, has gone on behaving like a Great Power long after the capabilities to perform such a role have disappeared.[2] One of the examples often cited is Britain's determination to continue to play a defence role on the world stage rather than concentrate sooner on a more realistic medium power role in Europe.

There is an element of truth in this thesis but no more than that. It is true that Britain has found (and continues to find) the task of adjusting to reduced circumstances difficult to deal with. There have certainly been times (like Suez in 1956) when policy-makers have overestimated Britain's power for independent action. It is also probably true that the desire to continue to play an important role outside Europe delayed the final decision in the 1960s to concentrate attention on the Continent.[3] Nevertheless, supporters of the 'Great Power illusion' thesis tend to confuse the rhetoric of British policy-makers with their actual motivation. Although there was a great deal of talk in the late 1940s and 1950s, in particular, about the need to continue Britain's Great Power status, it is clear from the records which have been made available in recent years that from the latter stages of the war onwards there was a clear recognition in Whitehall of Britain's reduced position in the world.[4] At the same time, there was also an awareness that Britain still retained important economic, political and strategic interests in different parts of the world which would still have to be defended for as long as posible. Britain's strength would have to be supplemented through alliances and adjustments would have to take place as colonies gained independence. But it was perceived that there was still value in continuing to play a role outside Europe.[5] The rhetoric of Great Power status helped in 'the game of bluff' which had always played a role in Britain's external relations, even in the days when 'the sun never set upon the British empire'.[6]

Whether this rhetoric actively prevented Britain from coming to terms with her European destiny is also open to question. From Ernest Bevin onwards Britain's attitude towards European cooperation was very different from many of the supporters of European integration, like Monnet and Schuman. The reason why Britain did not participate in the Messina talks or sign the Treaty of Rome was as much to do with the fact that she did not agree with the supranational aspects of the EEC as it was to do with Britain's determination to continue her world role. (And that world role was less to do with illusions of Great Power status than with the reality of defending concrete interests in a rapidly changing international environment.) Since Bevin's time Britain has accepted that it has important interests in Europe, especially in the security field.[7] But the legacy of empire has produced what have been perceived to be important, continuing interests outside Europe as well. Given Britain's dependence on world trade it is not surprising that she should have vital interests in international order and stability.

The result has been that Britain has faced a continuing dilemma on which to focus. Should priority be given to European defence or global security? What balance should be struck between these two (of the three) major circles of British policy identified by Winston Churchill? The attempt to deal with this dilemma has been a constant theme in British defence policy throughout the post-war period.

As the last chapter attempted to show, in the late 1940s the British Chiefs of Staff debated whether the Middle East or Europe should be given priority in strategic planning.[8] For a number of years two of the three Chiefs of Staff held out in favour of the primacy of the Middle East which was seen as the key base for an air war against the Soviet

Union and the protection of global British interests. Slowly but
surely, however, as Britain played an important role in the consolidation of Western European defence the emphasis began to switch away
from the Middle East towards Europe. This relative change in
strategic priorities was symbolised by the Defence Committee's decision, in March 1950, to send reinforcements to Europe in the event
of war and by the Paris Agreements of 1954.[9] Despite the fact that the
Paris Agreements contained an important 'let-out clause' allowing
Britain to withdraw forces from the Continent when her interests were
threatened overseas or when her economy was ailing, the commitment
of 70 000 troops to European defence inevitably impinged on Britain's freedom of manoeuvre overseas. This was particularly serious
because of the periodic crises Britain was facing throughout the fifties
and early sixties in places like Malaya, Suez, Kenya, Cyprus, Aden,
Kuwait and Borneo.

Two main problems confronted British policy-makers. Firstly,
there was the problem of manpower. Indian independence, as Philip
Darby has shown, deprived Britain of a large reservoir of forces which
had traditionally been used to garrison the empire.[10] The situation
was made worse in 1958 by the decision to abandon conscription by
1962 and rely on an all-volunteer army. In 1956, the Hull Committee
had concluded that 210 000 troops were required to fulfil Britain's
commitments.[11] In practice, however, the Army declined from around
400 000 in 1955–6 to around 180 000 in 1963 causing, what Richard
Rosecrance has described as 'a shortfall of critical importance'.[12]
Leaving NATO forces and the Strategic Reserve to one side, this left
only about 25 000 troops for worldwide commitments. Consequently,
if substantial forces were to be retained in Europe, and if simulta-
neously a series of international events required troops to be sent
overseas, which is precisely what happened in the late 1950s and early
1960s, then the result was bound to be the overstretching of Britain's
military resources. This is exactly what occurred in the early 1960s.[13]

The second problem was one of procurement and training. The
commitment to Europe meant that many of the weapons and much of
the equipment produced was designed specifically for defence against
the heavy armoured formations favoured by the Soviet Union and
Warsaw Pact. The tactics and training of BAOR and the Second
Tactical Air Force in Germany also had to be geared to the special
needs of continental defence. Although some of the weapons and
expertise could be used elsewhere, by and large the requirements of
defence on the Central Front and overseas military intervention were

very different. Given the perennial restrictions on the defence budget and the increasing expense of the major items of military hardware, there was an inevitable problem of providing the kind of air mobility and air portable equipment necessary for 'out-of-area' operations. Despite the development of concepts like the 'naval task force' and the 'central strategic reserve', the kind of capabilities required to provide effective implementation of these ideas were not forthcoming.[14] In the second half of the 1950s and the early 1960s this was very apparent in some overseas operations such as those conducted at Suez in 1956 and Kuwait in 1961.[15]

The dilemma for successive British governments has remained how to strike the right balance between European and global security interests. In attempting to resolve this dilemma there has been a tendency to shift the emphasis depending on where the greatest threats appeared to be. Having committed themselves to Europe in the mid-1950s, overseas operations nevertheless led to a scaling down of BAOR in the late 1950s and an increasing emphasis by the Minister of Defence, Harold Watkinson, on greater air-lift and air-portable equipment for 'out-of-area' contingencies. In 1962, the Chiefs of Staff consolidated the thinking of the previous few years by producing a Chiefs of Staff Paper which argued that war in Europe was increasingly less likely and that a major emphasis had to be placed on threats to Western interests in Africa and Asia.[16] As a result the focus of defence thinking formally shifted to the east of Suez region.

With the strain on Britain's armed forces in the early 1960s, the new Labour government under Harold Wilson undertook a major review of defence policy between 1964 and 1968. Although the Prime Minister himself was inclined to believe that 'a thousand men on the Himalaya's was preferable to a thousand men in Europe', the government finally concluded that the priority had to be given to European defence.[17] Faced with the perceived need to choose between Europe and the 'open-seas', the Wilson administration chose to reinforce their growing political and economic interests in Europe by concentrating Britain's defence resources on the Continent. In 1967, the decision was made by the Cabinet to withdraw British forces from east of Suez by the mid 1970s and subsequently, in 1968, after further financial difficulties, 'the retreat from empire' was to be further speeded up.[18]

The 1964–8 defence review was therefore a major milestone in British foreign and defence policy in the twentieth century. Faced with what the government of the day saw as a major choice between

European and global security interests, they concluded that Europe was more crucial to Britain's security interests than the remaining commitments overseas. In one sense this decision resolved a major problem which foreign and defence planners had wrestled with for many years. Europe could now become the unambiguous focus for defence planning. In another sense, however, the dilemma remained. Given that Britain retained important interests east of Suez and the United States in particular was increasingly keen for Britain to continue to play a role outside Europe, governments since the late 1960s have had to decide what residual capability to provide for 'out-of-area' operations. Should Britain have a dedicated 'special capability' designed for rapid intervention or a more 'general capability' which could be used in Europe and overseas if the occasion demanded. Given that this was not to be one of the four vital pillars of British defence policy (as identified by the 1974–5 defence review), how much (or how little) should be spent on such a force? Should it be only a token force for largely political purposes or a balanced, well-equipped force capable of effective military action?

Ever since the withdrawal from east of Suez in the early 1970s there has, in general, been a weakening of Britain's interventionary capabilities.[19] Some increased attention was given to overseas security with the return of the Conservative government under Mr Heath in 1970 but, to the regret of some Tory backbenchers, it did not add up to a significant reversal of the previous Labour government's priorities. Thus by the time John Nott undertook his defence review in 1981 Britain possessed only a very limited intervention capability indeed. In his White Paper, *The Way Forward*, Nott announced a number of measures designed to enhance the 'out-of-area' flexibility of Britain's armed forces.[20] These included a 'special equipment stockpile for limited operations overseas', the 'stretching of the RAF's thirty Hercules aircraft to increase their capabilities over short ranges by the equivalent of eight new aircraft, the ability to carry out the coordinated drop of a parachute assault force, improvements in command and control for overseas operations, and a headquarters (Eighth Field Force) specifically earmarked for such operations'.[21] Even with these improvements, as James Wyllie has argued, Britain's intervention capabilities remained very limited and it was 'obviously expected that any traditional intervention would be conducted in conjunction with allies'.[22]

Then came the Falklands War in the spring and early summer of 1982. As a result a new debate broke out over whether the government

had got the balance right. In its review of the campaign the government concluded that 'new and additional equipments were necessary
to increase the mobility, flexibility and readiness for operations within
the NATO area and elsewhere'.[23] The government made it clear that it
had not changed its mind about its major strategic priorities. 'It is still
in Europe that we and our allies face the greatest concentration of
Warsaw Pact forces.'[24] As such the four main priority areas were still
to have the first call on the resources available. Some enhancement of
intervention capabilities, however, was envisaged. To the combat
arms unit already in being (including two parachute battalions, an
infantry battalion and engineer support) were added an armoured
reconnaissance regiment, an artillery regiment, an Army Air Corps
Squadron and further logistic support units. In the words of a
government White Paper on Defence published in December 1982,
these forces: 'Taken together with the amphibious capability of the
3rd Commando Brigade gives us a greatly improved ability to
respond to the unforeseen in a flexible and rapid way.'[25] In addition,
the government announced that the two assault ships, HMS *Fearless*
and HMS *Intrepid*, which were to be phased out (under the 1981
review), would be retained.

These changes added up to a limited, but nevertheless significant
improvement in Britain's capability to perform 'out-of-area' operations and a recognition that what had previously been little more than
a token force should be upgraded into a properly constituted intervention force. What 'a properly constituted intervention force' should
consist of has remained a matter of some dispute. By definition
'unforeseen events' which the intervention capabilities are designed to
deal with are unpredictable and difficult to plan for. Trying to achieve
greater flexibility and mobility is sensible but given the restrictions on
the defence budget and given the government's priorities there is a
definite limit to how much mobility and flexibility can be achieved. A
limited force capable of meeting certain kinds of emergencies will not
be suitable for every eventuality which might require intervention.
Forces required for naval operations in the Gulf region will inevitably
be different from those required for internal security operations in
Hong Kong, or dealing with threats by Guatemala against Belize, or
disturbances in Gibraltar, or a renewed confrontation with Argentina
over the Falkland Islands.

Not surprisingly, there have been some critics who have argued that
the government has not gone far enough in restructuring Britain's
defence forces after the Falklands War. For those writers such as

John Wilkinson and Michael Chichester, Britain should adopt a new role in which Europe is given less priority and more attention is given to 'airmobile and amphibious intervention forces' for operations overseas.[26]

Linked closely to the need for better intervention capabilities, according to these critics, is the need to switch more resources towards the surface naval fleet. Given the flexibility of naval power more spending on surface ships, it is argued, would produce forces capable of operations not only in the Eastern Atlantic and Norwegian Sea but also east of Suez as well. Most members of the naval lobby emphasise Britain's role as a trading nation and the vital importance of freedom of navigation and international order.[27] In a world of growing turbulence and instability, strong and versatile naval forces are seen as providing a crucial element in Britain's continuing need to protect its worldwide commercial and political interests. For those who hold such views, the importance of intervention capabilities and naval power in the Falklands War and in the Gulf in the late 1980s provide eloquent testimony for their case.

Faced with arguments like these the government has found the task of striking the right balance between European and global security interests far from easy. Since the Falklands War government policy has been characterised by much more indecision than was evident in the 1981 review. Phrases like 'we would like to do more' (in terms of intervention capabilities) and 'we intend to retain a Royal Naval Force of *around* 50 frigates and destroyers' suggests that decisions have been postponed as the government attempts to keep its options open.[28] Whether such an approach is to be preferred to one which establishes clear priorities is a matter we will return to later.

3 Europe Versus the 'Special Relationship'

Despite the continuing problem of achieving exactly the right balance between European and overseas security, the major dilemma over which should have priority has clearly been resolved in Europe's favour. The debate about whether to emphasise closer defence collaboration with our European allies or whether to try to sustain a special defence relationship with the United States, however, remains largely unsettled in the late 1980s. Clearly the two areas are not mutually exclusive. It has been possible in the past (and remains so today) for Britain to forge close links with her continental allies while at the same time preserving a whole range of close ties in the defence field with the United States. As in other areas of defence policy, the key question has been one of priorities. Every government since 1945 has faced the problem of whether to emphasise 'the European circle' or 'the American circle'. Successive governments, both Labour and Conservative, have had to decide what balance to strike between these two key areas of British concern.

In the first twenty years after the Second World War the need to preserve a close, preferential relationship in the intelligence and wider defence fields with the United States was an overarching priority for Britain. As Chapter 2 has shown, Britain was prepared to play an important role in helping to coordinate Western European defence and also from 1948 onwards to commit herself to continental defence in peace and war.[1] With the movement towards European integration in the 1950s, however, Britain cited her overseas interests and close ties with the United States as an excuse for non-participation. The legacy of the unique wartime relationship with the United States, when Britain's very survival was at stake, had a profound influence on post-war officials and politicians – most of whom had seen at first hand during the war the value of close ties with the United States.[2] In line with Churchill's plea in his 'Iron Curtain' speech, in March 1946, for a continuing special military relationship 'between like-minded peoples', the post-war Labour government sought to 'entangle' the United States in British and Western European security interests.

Both the Foreign Secretary, Ernest Bevin and the Chiefs of Staff recognised that the range of British defence commitments could not be undertaken alone. British security in Western Europe and in the rest of the world could only be guaranteed in close alliance with the United States. In consequence, as Richard Best has shown, the immediate post-war years witnessed the gradual restoration of Service and intelligence links which formed the basis of an informal alliance between the defence establishments in the two countries.[3] It was on these foundations, despite some political and economic differences, which Bevin built a wider strategic relationship, initially in the Middle East and later in Western Europe with the establishment of NATO.

For the Labour government in the late 1940s the consolidation of European defence was a major interest. However, as Elizabeth Barker has pointed out, Britain's leading role in the establishment of the Western Union in 1948 was 'a sprat to catch a whale'.[4] Britain realised that if centuries of American foreign and defence policy were to be overturned and the USA was to be persuaded to commit itself to a peacetime alliance, Western Europe would have to demonstrate a willingness to defend itself. If that was to occur Britain would have to take a lead. Thus Britain's role in the creation of the Brussels Pact, in March 1948, had rather more to do with a desire to commit the United States to British (and continental) defence than to set up a Western European defence bloc which would stand on its own. 'Defence in depth' on the Continent was important to Britain but it was clear to British policy-makers that Western Europe could not defend itself without the power of the United States.[5] A close military relationship with the USA was the key to British security.

The aim of supplementing Britain's declining power through a close relationship with the United States also characterised British policy outside Europe. Despite periodic American criticisms of British imperial policy and British concern to limit American influence in certain areas of traditional interest, one of the hallmarks of the 'special relationship' which developed in the 1950s and early 1960s was a worldwide security partnership. British and American troops fought together (under the UN flag) in the Korean War between 1950 and 1953. American financial assistance helped Britain fight a protracted war in Malaya between 1948 and 1960. In the Jordan and Lebanon landings, in 1958, both countries provided each other with important assistance to sustain Western influence in two countries threatened by radical Arab nationalism. The establishment of SEATO and CENTO in the mid-1950s also helped to contribute to the protection of joint American and British interests. There were times when important clashes of interest did occur between the two states, particularly in the Suez crisis of 1956, but this only spurred British policy-makers to re-establish the 'special relationship' as quickly as possible. Suez apart, the contribution Britain made during this period to the global security partnership was welcomed by successive American administrations. By the early 1960s earlier criticism of British imperialism had long since disappeared. In December 1964, the American Secretary of Defense, Robert McNamara, emphasised to his British opposite number, Denis Healey, how important he believed Britain's role east of Suez was to the United States. Denis Healey told his colleagues in the Cabinet that what the Americans wanted Britain to do was 'to keep a foothold in Hong Kong, Malaya, the Gulf region, to enable us to do things for the alliance which they can't do'. The American government clearly believed at this time that British forces were 'much more useful to the Alliance outside Europe than in Germany'.[6] Sustaining the 'special relationship', in part at least, meant sustaining a British military presence east of Suez.

Another important area of cooperation which characterised the 'special relationship' in the 1950s and early 1960s was in the field of atomic energy. In the late 1940s, as Margaret Gowing has noted, atomic energy was the one major exception to the growing intimacy between the two countries.[7] The McMahon Act of 1946 which cut off nuclear cooperation with Great Britain was a major setback to the Labour government of the day. Attlee and Bevin expected the United States to honour the wartime agreements, made at Quebec in 1943 and Hyde Park in 1944, which promised to continue the nuclear cooperation between the two countries established in the wartime Manhattan project into the post-war period. Roosevelt's death, a lack of awareness of the secret wartime agreements in the new Truman administration, and the perception in Congress that nuclear weapons were 'a sacred trust', all contributed to the post-war legislation to preserve an American nuclear monopoly.

With the American door closed, one option for Britain might have been to seek collaboration with other Western European countries with expertise in the nuclear field.[8] Britain's reaction to the disappointment of the McMahon Act was to pursue an independent nuclear programme while at the same time attempting to break down the political barriers to cooperation with the United States. The aim was to show the United States that Britain was determined to become a nuclear power, with or without American help, and that Britain had a contribution to make to the American nuclear programme. This was a strategy which proved to be very successful indeed. In January 1948, the log-jam of non-cooperation began to break with a *modus vivendi* agreed between the two governments which permitted a limited amount of nuclear collaboration to take place. More importantly, in 1954, the Atomic Energy Act was modified to allow collaboration in a number of important fields. This was followed in 1958 by the repeal of the McMahon Act and a new Atomic Energy agreement between Britain and the United States which provided for wide-ranging nuclear cooperation. From this agreement (as amended in 1959) Britain was to acquire a preferential relationship with the United States, which no other American ally then or since has been able to achieve. These nuclear exchange agreements together with the Thor Agreement of 1958, the Skybolt Agreement of 1960 and the Polaris Agreement of 1962 were to bring nuclear weapons to the very heart of the 'special relationship'.[9]

By the early 1960s, therefore, despite the importance of Europe in British military thinking, the 'special relationship' with the United States was the key focus. At the same time as the Polaris Sales Agreement was being signed with the United States, events were occurring which increased the importance of the European circle. The first (unsuccessful) application to join the Common Market made in 1962 signalled a change in British attitudes towards the Community. Harold Wilson applied again (unsuccessfully) in 1967 at a time when his government was undertaking a major defence review which resulted in British troops being withdrawn from east of Suez and greater concentration being given to European defence. As a result, the global security partnership which had previously been such an important part of the 'special relationship' was irretrievably weakened just when the United States sought British assistance in the Vietnam War. Symbolically the Prime Minister also chose to talk of the 'close' rather than the 'special' relationship with the United States. Moreover, the personal warmth which had characterised the Roosevelt–Churchill and Kennedy–Macmillan relationships was clearly absent from the Johnson–Wilson relationship.[10]

The refocusing of British foreign and defence policy towards Europe was taken a stage further by the Heath government in the early 1970s which saw the third and successful application to join the EEC. As Henry Kissinger relates in his memoirs: 'Heath dealt with us with an unsentimentality quite at variance with the special relationship. The intimate consultation by which American and British policies had been coordinated in the postwar period was reduced to formal diplomatic exchanges.'[11] Heath's concern to symbolise the change in British policy towards Europe caused him deliberately to play down the 'special relationship' – or what he preferred to call 'the natural relationship'.[12]

In the defence field this reorientation of policy caused Heath to join with other European countries in refusing to allow American planes to use British bases to reinforce the Israelis in the Arab–Israeli War of 1973. Increasing attention was also given to a range of joint procurement projects with Britain's European allies to supplement the growing political and economic ties with the Continent. At the nuclear level (the core of the 'special relationship') some consideration was even given to the possibility of collaboration with the French on the next generation of strategic deterrent to replace Polaris.[13]

It should not be inferred from this growing Europeanism in British defence policy, however, that either the Wilson or the Heath governments were determined to end the close defence relationship with the United States. What occurred in the late 1960s and early 1970s was a decision to increase the emphasis on European cooperation in the defence field (as well as the political and economic fields). The 'special relationship' was undoubtedly weakened as a result but a whole range of intimate and important defence links remained. Despite Wilson's threat to renegotiate the Nassau Agreement with the United States, his government was responsible for the smooth deployment of the Polaris force in the late 1960s and the continuation of close working relationships between British and American scientists in a wide area of defence projects. The Heath government also finally concluded that the Anglo-American nuclear relationship had far more to offer than collaboration with the French. In April 1973, the decision was made to initiate the Polaris Improvement Programme with the help of the United States. For the Heath government the balance had swung more towards Europe but 'the natural relationship' with the United States nevertheless remained of major importance in the defence field as in other fields. What occurred was a change of emphasis.

If Heath had allowed 'the rope to go slack' to a certain extent, the Wilson and especially the Callaghan governments from 1974 to 1979 attempted to inject a little more warmth into the relationship at the highest level as the focus shifted back somewhat towards close ties with the United States. James Callaghan, in particular, got on well with Henry Kissinger and later President Carter and refused to pursue the distinctively European policies characteristic of the previous Conservative administration. The formal diplomatic approach to relations with the United States was dropped in favour of a return to the more intimate consultations which had traditionally characterised the 'special relationship'. In the nuclear field the decision was taken to upgrade the Polaris force through the secret 'Chevaline' project which, as Francis Pym was to reveal to the House of Commons in 1980, involved a close working relationship with the United States. By the time the Callaghan government lost office in 1979, despite the public declaration that it had no intention of producing a new generation of nuclear weapons, a number of studies had been produced on a successor to Polaris. These included a study on the purchase of Trident missiles from the United States.[14]

Important as the residual 'special relationship' was to the Callaghan government, there was no return to the kind of wide-ranging intimate partnership which had prevailed in the 1950s and early 1960s. The Wilson and Heath periods had left their mark. In particular Britain no longer made a significant contribution to extra-European defence. The continuing decline of British military power was highlighted by the 1974–5 defence review which initiated a contraction in Britain's contribution even to European defence. The next two or three years also saw frequent incremental cuts in British defence spending which brought loud criticisms from American military circles.[15] Even with the Callaghan government's reversal of the downward spiral in 1977 and its commitment to an annual 3 per cent real increase in defence spending (along with the other NATO allies), there is no doubt that by the end of the late 1970s Britain's contribution to the special defence relationship was considerably less than it had been in the 1950s and early 1960s.

One of the objectives of the Thatcher government which came to power in May 1979 was to balance more effectively the European and Atlanticist strands of British foreign and defence policy. The two were seen as complementary rather than in opposition. As Mrs Thatcher fought to renegotiate Britain's contribution to the Community budget, British policy tended to be characterised by a coolness towards the EEC. In contrast, the relationship she established with the American President, Ronald Reagan, was warm and friendly, re-establishing the close personal ties at the highest level of government which had been a hallmark of the 'special relationship' in the past. After the semantic downgrading of the label 'special' by Wilson and

Heath with their use of the terms 'the close' and 'the natural' relationship, Mrs Thatcher went out of her way in December 1979 to refer to 'the extraordinary relationship'. In the words of the Oxford dictionary 'extraordinary' means 'out of the ordinary, special'.[16]

In the defence field the importance Mrs Thatcher attached to the continuation of close ties with the United States was symbolised by the Trident agreements of 1980 and 1982.[17] Shortly after taking office Mrs Thatcher moved swiftly to decide on the Trident C4 missile as a successor to Britain's Polaris force in the 1990s. In October 1981, President Reagan decided to speed up the Trident D5 programme as part of the American military build-up which he had promised in his election campaign. This left Britain in a difficult position. The C4 missile which Britain had agreed to purchase was perfectly adequate for Britain's needs. The United States, however, was now planning to phase out the C4 system earlier than had originally been decided. As Britain was not planning to deploy the Trident force before the early 1990s this would have left her with what the government itself described as a problem of 'uniqueness': the difficulty of continuing to operate the system on her own without American assistance.[18] This, almost certainly, would have resulted in a number of logistic, operational and financial problems. As a result the government decided in March 1982 that the benefits of 'commonality' indicated that Britain would be better served by purchasing Trident D5 missiles rather than the C4 missiles.

Apart from the political symbolism of the Trident agreements, Mrs Thatcher's commitment to strengthening the Anglo-American alliance was shown in other ways, particularly the strong diplomatic support given to the United States as international instability increased in the early 1980s. She provided strong backing for the United States in its campaign against Soviet intervention of Afghanistan. She also played a key role in the implementation of NATO's dual-track decision in December 1979 on intermediate range nuclear forces (INF).[19] After the failure of the negotiations with the Soviet Union between 1981 and 1983 the British government stuck firmly to its commitment to deploy cruise missiles in Britain, despite considerable domestic opposition. The British government also provided diplomatic stiffening to some of its Western European allies who appeared to waver in their commitments.[20] Neither was the relationship one-sided. Important assistance was given to Britain by the United States during the Falklands War in 1982.[21] Indeed, such was the perceived significance of that assistance that the American Secretary of Defense, Caspar Weinberger, was given a knighthood by the British government (after his retirement) in early 1988. In line with the reciprocal nature of the renewed security relationship between the two countries, Mrs Thatcher allowed American aircraft to fly from British bases to attack Libya in April 1986 (in retaliation for Colonel Gaddafi's support for international terrorism) – even though other Western European states refused assistance.[22] In similar vein, when the United States found itself alone in its attempt to provide protection for Kuwaiti ships threatened by Iran in the summer of 1987, Britain was the first of the Western allies to commit naval forces to the Gulf region in support of the United States.[23]

Despite this renewal of the 'special relationship' by the Thatcher government in the 1980s, however, the European dimension of British defence policy was not wholly neglected. The attempt to balance European cooperation with close Anglo-American relations led Mrs Thatcher, on a number of occasions, to side with her European allies in disputes with the United States. The most notable of these centred on the question of sanctions against the Soviet Union after martial law was declared in Poland, in 1981. Mrs Thatcher supported her European allies in refusing to introduce sanctions and as a result British as well as other European firms for a time were black-listed by the United States.

Despite occasional clashes, the main thrust of British policy has been to act as a bridge between the United States and Europe. Britain has attempted to explain American policies to her European allies and to transmit European anxieties to the United States over such issues as trade policy, arms control, relations with the Soviet Union and Star Wars technology.

Britain's approach to the American Strategic Defense Initiative (SDI), in particular, is a good illustration of the policy of balancing Europe and the United States. When President Reagan announced the SDI programme in March 1983 Britain, like her European allies, was taken by surprise. Despite the significance of the announcement for Western strategy no consultations had taken place with the European states. After a period of silence while the British government considered the implications of SDI, Mrs Thatcher indicated her general support for the programme and sought to encourage British firms to participate. Behind this desire to collaborate with the US, however, was a deep anxiety shared by most of Britain's European allies. When Mrs Thatcher went to Camp David, in December 1984, her endorsement of President Reagan's initiative was shown to be conditional. Britain, she said, would support SDI as a research project, provided four main conditions were met. Firstly, that the United States' and Western aim should be to achieve a balance with, rather than superiority over, the Soviet Union. Secondly, that SDI-related developments would, in view of treaty obligations, have to be a matter of negotiations with the Soviet Union. Thirdly, that the overall aim should be to enhance, rather than undermine, deterrence. And fourthly, that East–West negotiations should try to achieve security with reduced levels of offensive systems on both sides.[24]

This qualified British support for the SDI programme was spelled out even more clearly in Sir Geoffrey Howe's speech to the Royal United Services Institute for Defence Studies in March 1985.[25] Although the Foreign Secretary attempted a balanced assessment of SDI, providing some support as well as expressing some scepticism, it was the scepticism which stood out. The speech raised questions about the cost of the programme and the counter-measures it might provoke from the Soviet Union, as well as its impact on arms control and the credibility of the US commitment to Europe. The last of these worries was to be found in all West European capitals at this time.

Consequently, although Britain took the lead and signed an agreement with the United States in 1985 allowing British companies and universities to participate in the SDI research programme, British policy-makers were nevertheless concerned to pass on their anxieties to America which they shared with other West European states. Similarly, since that time, as various agencies within the American administration have debated whether to move from a 'narrow interpretation' to a 'broader interpretation' of the 1972 ABM Treaty which, it is claimed, would allow research outside the laboratory in space, Britain has continued to urge restraint on the American government. Recognising growing support within the US administration in the mid-1980s for the need to test various SDI technologies more effectively, Mrs Thatcher modified her position somewhat in 1987 and indicated that Britain accepted that the definition of a laboratory, 'as being a room with four walls', might be too restrictive. She recognised the need for research to discover the feasibility of the project. Beyond this limited flexibility, Britain has continued to argue forcefully against those, especially in the Pentagon, who have tried to argue in favour of a limited deployment of SDI in the near future. In so doing Britain has taken the lead in expressing the anxieties of other West European states that the 'broader interpretation' of the ABM Treaty could undermine progress towards better relations with the

Soviet Union and the prospect for future arms control agreements.[26]

This coincidence of interest between Britain and Western Europe has not been confined to the SDI question. The INF negotiations and the Global Double Zero Agreement reached in December 1987 has also brought about an increased level of consultation between London, Paris and Bonn. Although Mrs Thatcher supported the American desire to achieve an INF agreement based on the zero-zero formula and indeed actively encouraged the wavering West German government of Chancellor Kohl to give its support, she was conscious of the political implications of the agreement for Western Europe. For most of the West European states a de-emphasis of theatre nuclear weapons has the effect of helping to decouple the United States from Western European security. It was also seen as giving added significance to the superiority of Soviet conventional and chemical forces in Europe. As a result, in a speech she gave in Bonn, in late September 1987, she argued against too much euphoria over the Global Double Zero Agreement and pinpointed the difficulties which the agreement would create for Western European defence if Soviet advantages were not addressed.[27]

Faced with an American determination to conclude an INF agreement with the Soviet Union despite residual European anxieties, France and Germany moved swiftly in the autumn of 1987 to upgrade their joint military collaboration by setting up a mixed Franco-German brigade and a Defence Council which they hoped would form the basis for wider European cooperation in defence. Despite the military problems associated with Chancellor Kohl's idea of a joint brigade, the force was symbolic of a deepening worry in Western Europe in the latter half of 1987 that the US was likely to reduce its involvement in European defence in the future. As a result an increasing interest in European defence cooperation was discernible in a number of Western European states.

In the past numerous attempts have been made to encourage the development of greater European defence cooperation. The results so far have been rather limited. This has been partly because of the reluctance of both Britain and Germany to weaken their defence links with the United States. With West Germany beginning to consider its military relationship with France more seriously much will depend in future on the role of Britain.

Britain's approach to greater European defence collaboration, in the past, has been ambivalent to say the least. Despite the British role in promoting the Eurogroup and the Independent European Programme Group (IEPG) in the 1960s and 1970s and her involvement in a number of joint procurement projects (including Jaguar, the FH 70 and Tornado), there has been a reluctance to support efforts to create a broader European defence pillar. This has been largely the result of a long-standing British concern that greater European defence cooperation might be perceived in the United States either as evidence that the Western Europeans were now capable of providing for their own defence unaided, or as evidence that Western Europe was adopting a more independent line. Either way the United States might be encouraged to reduce its commitment to European defence. As a result of this anxiety Britain has traditionally attempted to walk a delicate tightrope between demonstrating to the US that the Europeans were prepared to collaborate more effectively together for the common good, while at the same time showing that such collaboration could not be sufficiently effective to allow the Europeans to stand on their own unassisted.

In recent years, this ambivalence towards European defence cooperation has become even more acute than in the past. American domestic difficulties (especially those resulting from the high budget deficit), together with a growing tendency to look towards Central America and the Pacific, suggests the possibility of some future American disengagement from European defence. Greater Franco-German cooperation also poses the risk that Britain will be left behind if she does not participate in this new wave of European collaboration. At the same time, the British government does not want to do anything which will encourage or accelerate the process of American disengagement. The nature of the British dilemma can be seen notably in the government's response to French attempts during the 1980s to revamp the Western European Union (WEU). Britain has been prepared to go along with the Mitterrand government's plan to enhance the status of the WEU but has remained sceptical about the contribution the WEU can, or indeed ought to, make to greater European defence cooperation. The WEU has been viewed in Britain as a body outside NATO which provides a useful forum for discussion but which ought not to be the main focus of greater harmonisation of Western European defence efforts.[28]

Despite the government's claim that there is no contradiction between the European and transatlantic dimensions of British defence policy, there have been occasions when conflicts between the two have occurred. The most dramatic example of this was the Westland affair in late 1985 and early 1986. This centred on a debate in the Cabinet over whether to rescue the ailing Westland helicopter company through an arrangement with a European consortium or the American company, Sikorsky. As far as the Defence Secretary, Michael Heseltine, was concerned the government had committed itself to greater European cooperation in the defence field. In 1978, a declaration of principles had been agreed between France, Germany, Italy and the United Kingdom in which every country accepted the need to foster greater European cooperation in the production of helicopters. In 1984, Michael Heseltine himself had been involved in a campaign to reinvigorate the IEPG and encourage wider European collaboration across the board on defence projects. In November 1984, the IEPG met at the full Defence Minister level and agreed on a number of potentially far-reaching measures to encourage the development of a European pillar in the procurement field within the Atlantic Alliance framework. In the words of the 1985 Defence White Paper:

> The European effort ... is not an alternative to transatlantic cooperation: on the contrary, a stronger and more cohesive European industry will contribute to the strength of the alliance as a whole and enable Europe to cooperate more effectively on level terms with the United States. The United Kingdom therefore intends to press ahead in this field with vigour and determination.[29]

Given this commitment to a more vigorous approach to European defence cooperation, it was not surprising that Michael Heseltine would react to the growing financial crisis in the Westland helicopter company by encouraging the formation of an Anglo-European consortium to make a rescue bid. At the same time, however, the Secretary of State for Trade and Industry, Leon Brittan, favoured a private enterprise alternative (to the European consortium of state-dominated firms) offered by the American Sikorsky company. The result was a major row within the British government which eventually led to the resignation of both Cabinet Ministers concerned and the most serious crisis in Mrs Thatcher's second term in office. The board of Westland's eventually came down in favour of the Sikorsky bid rather than that of the European consortium.

Despite the government's backing of the 1978 agreement and the principles on greater European cooperation agreed by the IEPG Defence Ministers in November 1984, Mrs Thatcher showed no inclination during the crisis to support Michael Heseltine's bid for the European option. Indeed in Mr Heseltine's detailed statement after his resignation in January 1986 he complained bitterly of the way the issue had been handled by the Prime Minister. Mrs Thatcher herself appears to have favoured the American option.[30]

The government's dilemma in trying to sustain the 'special relationship' and at the same time encourage greater European cooperation has also been illustrated by Anglo-French negotiations in late 1987 and early 1988. In the changing environment created by the INF agreement in December 1987 the British Defence Secretary, George Younger and his French counterpart Mons. Andre Giraud initiated a series of discussions designed to achieve greater collaboration between the two countries in various defence fields. A number of studies were set in motion to consider the feasibility of an Anglo-French nuclear armed stand-off weapon, possibly utilizing cruise missile technology. Talks were also reported to be taking place on the sensitive questions of nuclear targeting and submarine patrolling.[31] After further talks between the two Defence Ministers in early 1988 it was announced that agreements had been reached between the two governments to allow British troops to train in France for the first time since 1966 and to reopen transit and supply routes through France for British forces in the event of war.[32]

Even though the range of Anglo-French cooperation was limited (in many cases merely consisting of joint studies) the British government remained sensitive to American reaction to these developments. When American government sources expressed reservations about Anglo-French nuclear cooperation the British government argued that they would welcome tripartite collaboration which would include the United States.[33] There was also speculation in the press that the purpose of Anglo-French cooperation was designed to bring France closer to the Western Alliance. *The Independent* in particular expressed the hope that an agreement reopening logistic routes through France would be followed up with a similar agreement with the United States.[34]

The evidence suggests that the dilemma between European cooperation and the 'special relationship' remains an important one for defence planners in the late 1980s and may well become sharper over the next decade. The government has persisted with its policy of trying to balance these two important areas of defence policy. The task of complementing these two defence interests, however, is far from easy. Whether the government can or should attempt to keep its options open, as it has in the past, is a question we will return to in Chapters 9 and 10.

4 Nuclear Weapons Versus Conventional Forces

Britain's determination to maintain a 'special relationship' with the United States for most of the period since 1945 has been closely tied to the determination of both Labour and Conservative governments to become, and then remain, a nuclear power. As Chapter 3 has shown, the Attlee government in 1945 was reluctant to accept an American monopoly of nuclear weapons and decided in 1947 to produce a nuclear capability, partly to facilitate British independence, and partly to encourage the restoration of interdependence with the United States at some point in the future. This latter objective having been achieved in 1958, independence and interdependence have remained important themes in British nuclear policy ever since. From the late 1950s on, Britain has sought to maintain a sophisticated nuclear capability under national control through reliance on American technology. At the same time, however, successive governments have attempted to retain a spectrum of balanced, well-trained and well-equipped conventional forces in order to perform a wide range of alliance and national tasks which could not be met with nuclear weapons alone. Given the economic constraints on the defence budget and the effects of defence inflation (which has consistently outrun 'ordinary' inflation), the result has been a continuing struggle over whether to give priority to nuclear or conventional forces.

From the latter stages of the Second World War, despite the fact that Britain was not to have a nuclear capability of its own until the mid-1950s, strategic planners recognised the value of atomic weapons.[1] As early as 1946 the Chiefs of Staff were arguing that in the nuclear age the best method of defence against atomic bombs was likely to be the deterrent effect of retaliation.[2] The threat of an atomic offensive was seen as the most effective way of deterring an aggressor and if deterrence failed such an offensive was seen as the best way to limit damage to the United Kingdom.[3] As a result of this stress on the importance of nuclear weapons, the decision was taken in January 1947 to produce an independent nuclear deterrent and by the late 1940s the programme was given super priority status by the government.[4] The need for such an emphasis on the nuclear component of defence policy, however, was not acceptable to all officials involved in the defence planning process. Sir Henry Tizard, the Chairman of the Defence Research Policy Committee and the Chief Scientific Adviser to the Minister of Defence, fought a sustained campaign during 1949 and 1950 to try to downgrade the emphasis on the nuclear weapons programme. Tizard argued that the priority accorded to atomic weapons was affecting the ability to provide a range of conventional weapons needed for essential defence tasks. In particular, he pointed to the vital need for Britain to defend itself more effectively through the provision of guided weapons. Tizard was opposed by Lord Portal, who presided over the Atomic Energy Council within the Ministry of Supply. Portal, who was a key figure in the nuclear programme, was committed to retaining the top priority accorded to atomic energy and he used his former citadel, the Chiefs of Staff Committee, to pursue his campaign against Tizard. When the Defence Committee came to consider the two positions, in April 1950, they found the task of adjudicating between the nuclear and non-nuclear programmes very difficult indeed. As a result they produced a compromise solution in which both guided weapons and the atomic weapons project received joint overriding priority![5]

Disputes over the right balance between nuclear and conventional weapons were to become a major cause of strain within the Chiefs of Staff Committee during the 1950s. In the Global Strategy Paper produced in the summer of 1952, the COS had agreed that Britain needed sufficient conventional and nuclear forces to do four things: 'to exercise influence on Cold War policy, to meet NATO obligations, to prepare for war in case the deterrent failed, and to play a part, albeit a small one, in the main deterrent, the air offensive'.[6] In essence the Global Strategy Paper represented a compromise between the three Services, especially between the Air Force idea of the importance of nuclear weapons and the short war concept, and the Navy's emphasis on sea communications and what was described as the 'broken-backed war' concept.

With the pressure for economic reform in 1953, and the 'Radical Review' which was initiated, the tensions between the Services and between nuclear and conventional weapons priorities increased. For the Chief of the Air Staff, Sir John Slessor, the essential priority had to be deterrence through preparations for a strategic air offensive using nuclear weapons. For the Air Force, the first six weeks of any future war would be decisive. Although the First Sea Lord, Sir

Rhoderick McGrigor, also sought a role for the Navy in the strategic
air offensive (from naval carriers) he was not convinced by the
argument that a future war would be of short duration. War, he
argued, would be likely to last much longer than six weeks and as a
result there was a major requirement for naval forces to operate in a
protracted campaign.[7]

Despite the institutional and doctrinal differences within the Chiefs
of Staff Committee, the Chiefs themselves were eventually prepared to
compromise over the balance to be struck between nuclear and
conventional forces in order to provide a united front to the government which was seeking economies in defence. The Conservative
government's campaign to save money led to a continuing search for
priorities and nuclear weapons were regarded increasingly as a
relatively inexpensive means to enhance Britain's security. This tendency to put even more emphasis on nuclear weapons increased after
the decision to develop hydrogen weapons in June 1954. Duncan
Sandys, the Minister of Supply, was particularly critical of the Navy's
carrier programme and led a campaign in favour of focusing more
attention on the development of Britain's nuclear deterrent. The
appointment of Sandys as Minister of Defence after the humiliation
of Suez, gave him the opportunity of shifting the balance even further
towards nuclear weapons – perceived as a useful way of enhancing
British prestige at a difficult time for the government and for Britain's
status in the world. It must be said, however, that the Sandys White
Paper of 1957, which put such stress on the future development of
nuclear capabilities, was not a radical departure from previous
policies. It simply confirmed and extended existing trends in British
defence policy. But in so doing it did put nuclear deterrence very
much at the centre of the strategic stage in Britain. In Richard
Rosecrance's words, 'it was not until 1957 that independent strategic
weapons were viewed as all important'.[8]

The pursuit of such nuclear priorities clearly had implications for
other areas of British defence policy, especially as Sandys also decided
to end national conscription. Conventional forces for operations in
Europe and outside were recognised as being important in the Sandys
White Paper. Nuclear weapons, however, were viewed as the main
priority. Consequently, as costs for research and development in the
nuclear field escalated in the late 1950s and early 1960s, especially as a
result of the Blue Streak project, conventional capabilities suffered.
Britain's ability to project military power overseas in particular
declined.[9]

Sandys' successor as Defence Minister, Harold Watkinson, attempted to redress the balance to a certain extent by improving Britain's Cold War forces. The commitment to the nuclear deterrent contributed to the inadequacy of Britain's conventional forces as the government attempted to cope with the range of crises which emerged in the early 1960s. As Britain struggled to stay in the nuclear race, following the cancellation first of Blue Streak in 1960 and then Skybolt in 1962, the conventional forces were put under increasing strain. By 1964, when the Wilson government took over, that pressure (created by Sandys' deliberate attempt to cut back on conventional capabilities but not on commitments) became intolerable.[10]

The series of reviews between 1964 and 1968 attempted to address the problem. Despite its criticisms in opposition of the independent nuclear deterrent, the Labour Party in office refused to downgrade the continuing commitment to the nuclear weapons programme in any significant way. Nassau was 'renegotiated' and the government decided to build four rather than five Polaris submarines. In essence, the Wilson government put into effect the nuclear programme inherited from its predecessors. With nuclear weapons retaining their central role, attention in the defence reviews shifted to balancing conventional capabilities and commitments. In an attempt to bring these closer into line British forces were withdrawn from the Far East and the Gulf region. With commitments reduced, further cuts in the capabilities to perform those overseas responsibilities could also be introduced. To meet a fixed defence budget ceiling the TSR2 aircraft was cancelled; no 'special capability' was to be retained for operations east of Suez; and the government decided eventually not to go ahead with the purchase of fifty F111 aircraft from the United States – initially viewed as a replacement for the TSR2. The decision was also made to phase out the Navy's carriers by the early 1970s.[11]

For a time in the 1970s it looked as if the traditional commitment to the nuclear deterrent force was beginning to waver. The Heath government postponed a decision to replace Polaris in 1973 and the succeeding Labour government, as part of its defence review in 1974–5, announced that it would not purchase a new generation of nuclear weapons after Polaris. In practice, however, very little changed. The Heath government initiated a programme designed to update the Polaris system and the Labour government committed itself to the secret 'Chevaline' project which eventually was to cost in the region of £1000 million.[12] At the same time, there was little attempt to improve Britain's conventional capabilities; indeed, quite the reverse. In the

1974–5 defence review major procurement projects were stretched, the remaining forces east of Suez were withdrawn or reduced in size, and reductions were ordered in specialist reinforcement capabilities.[13] A number of incremental cuts in 1975 and 1976 further weakened Britain's conventional forces.

This erosion of Britain's conventional capabilities in the mid-1970s arose even though the government did attempt to balance nuclear and conventional priorities. The 1974–5 review identified four key areas of defence policy: the nuclear deterrent, defence of the Central Front, defence of the Home Base, and defence of the Eastern Atlantic and Channel. Conventional forces were nevertheless hit hardest by the government's decision to reduce defence spending from 5.5 to 4.5 per cent of GNP. By the late 1970s the growing criticisms of Britain's weakening conventional capabilities led the government to try to re-establish a more effective balance between nuclear and conventional forces. The Callaghan government committed itself to the NATO plan to increase defence spending by 3 per cent in real terms from 1977 onwards. It also supported the Long Term Defence Programme (LTDP) designed, in particular, to achieve improvements in ten areas of largely conventional capability.[14]

The big question mark at the end of the 1970s was over the future of the British nuclear force. The Labour government had continued throughout the decade to maintain that it would not replace the Polaris system. At the same time, in secret, the 'Chevaline' programme continued and the government set up a review of possible alternative successors.[15] The momentum for the continuation of the Polaris programme into the early 1990s had clearly been established. Despite its public pronouncements the Labour government was prepared at least to consider the case for replacement. Whether it would have actually taken the decision to replace Polaris is impossible to say. In May 1979, Mr Callaghan was defeated by Mrs Thatcher and a new government came to power committed to maintaining the nuclear force and replacing Polaris.

The Conservative government was not only pledged to the strategic nuclear deterrent but was also determined to upgrade Britain's defences across the board. It soon discovered, however, that the task of balancing conventional and nuclear forces was more difficult than it seemed, even though the government was prepared to continue the real increases in defence spending initiated by the previous Labour government. The 1981 defence review represented a recognition by the government that the existing defence programme could not be implemented without a substantial increase in defence spending over and above the existing 3 per cent real increases.[16]

As we have seen the Nott review undertook to scrutinise every aspect of defence spending. Nothing was to be sacrosanct. In practice the decision to continue the nuclear deterrent had largely been pre-empted by the Trident Agreement of 1980. It would have been very surprising if the government's decision in favour of Trident in 1980, had been reversed less than a year later by the Nott review. After considering the arguments the government decided that the nuclear deterrent was a unique and crucial component of Britain's defence effort. In the words of the White Paper, *The Way Forward*, 'no other alternative application of defence resources could approach this in real deterrence insurance'.[17] The priority given to nuclear weapons ever since the late 1940s was to be maintained.

As a result cuts had to be found in Britain's conventional forces and Chapter 1 has shown that the government decided in favour of reductions in the Navy's surface fleet. Following the Falklands War, however, the debate over the priorities between different conventional capabilities was to resume.[18] What is interesting about the debate which has continued since 1982 (and which is still not concluded in 1988) is that, despite differences between the Services over which conventional capabilities should be cut, there has been very little disagreement over the commitment to continue Britain's nuclear force into the 1990s and beyond. Supporters of both the continental commitment and the naval lobby largely agree that the expenditure on Trident is essential.[19]

The one exception to this consensus has been Lord Carver who held the post of Chief of the Defence Staff between 1973 and 1976 (when there does appear to have been some wavering over the priority to be given to the nuclear deterrent in both the Heath and Wilson governments). In his writings Lord Carver has argued forcefully that 'the essential ingredient' in preventing the recurrence of war in Europe 'is the presence of US armed forces in the Western half of it'.[20] It is this presence which integrates American power in all its forms into the defence of Western Europe. According to Lord Carver, the significance of the nuclear forces of both the USA and the Soviet Union lies only in mutual deterrence and, if ever this should break down, in preventing each from using nuclear weapons against the other. He argues that 'as long as both are clearly and physically linked to the security of Europe ... this mutual deterrence against war keeps Europe at peace, rigid, anomalous and unpleasant as it may be for many people'.[21] The conclusion that Lord Carver has come to from this analysis is that there is no reason why European countries, including Britain, should maintain their own nuclear weapons. 'Britain,' he says, 'should waste no more money on nuclear weapons.'[22] It would be far better to spend the money on extra conventional capabilities, particularly the provision of 300 more tanks on the Central Front. Lord Carver's case against the British nuclear deterrent was not only that he saw it as wholly superfluous to the American nuclear armoury, but also that the force was lacking in credibility as a deterrent and dangerous to British security interests. In a speech he made to the House of Lords, in December 1979, he went out of his way to emphasise that his experience at the highest levels of the military establishment had convinced him of the futility of maintaining an independent nuclear deterrent force. He argued that:

> Over the years the arguments have shifted and I have read them all; but in that time I have never heard or read a scenario which I would consider it right or reasonable for the Prime Minister or government of this country to order the firing of our independent strategic force at a time when the Americans were not prepared to fire theirs – certainly not before Russian nuclear weapons had landed in this country. And again, if they had already landed would it be right and reasonable? All it would do would be to invite further retaliation.[23]

In terms of the public record Lord Carver appears to have been the only Chief of Staff or Chief of the Defence Staff in the whole of the post-war period who has taken this view. Many of them may have had their doubts about nuclear deterrence (including Lord Mountbatten) but Lord Carver has been the only one to advocate an end to the British nuclear deterrent.[24] Despite their major differences on resource allocation between the different conventional weapons projects, there would seem to have been a remarkable consensus within the Chiefs of Staff Committee in favour of a continuing emphasis on nuclear weapons. When cuts have had to be made the Chiefs have argued, often fiercely, about the priorities which should be given to particular conventional options but the nuclear deterrent appears never to have come under concerted attack within the Chiefs of Staff Committee. Despite the 'opportunity costs' for conventional capabilities associated with this continuing priority, the nuclear deterrent has continued to be viewed from the 1940s to the late 1980s as a 'crucial strategic element' of British defence policy.

Had the Labour Party won the General Elections in 1979, 1983 and 1987 this priority might have changed. In both 1983 and 1987, in particular, the party was led by two confirmed unilateralists who were committed to phasing out Polaris and cancelling Trident.[25] Whether they would have carried out the fundamental change in defence priorities contained in their manifestos is a question which can never be answered. As we have seen the Labour Party came to power in 1964 with what appeared to many to be a policy of abandoning the independent deterrent but failed to do so in practice.[26] Similarly, in the 1970s a Labour government announced its intention of not replacing Polaris but nevertheless continued the improvement programme and was actively considering a replacement when it lost power.[27] In the aftermath of their third electoral defeat pressures have arisen in the party to review the commitment to unilateralism before the next election in 1991–2. For the moment, however, the party remains committed to a non-nuclear defence policy.[28]

By the time the next election takes place the Trident programme will be well entrenched. More than half the £9 billion will have been spent and more again will have been committed. Such expenditure by the Thatcher government, as many commentators have pointed out, is likely to have a detrimental effect on Britain's conventional capabilities. Irrespective of whether a future Labour government decommissions the Trident force, a large amount of money will have been spent and conventional forces will have suffered. Although the figure of 3 to 6 per cent of the defence budget spent on nuclear weapons over a fifteen to twenty-year period does not appear to be a significant amount, the continuing expenditure on Polaris (of around 2 per cent of the defence budget) has to be included. Of even more importance, the expenditure on Trident represents around 8 to 10 per cent of the equipment budget and as much as 16 per cent of the *new* equipment budget. This is occurring in a period when the government is planning new tanks for the Army, new aircraft for the Royal Air Force and new ships for the Navy. At a time when the defence budget is declining in real terms, the government is attempting to retain a balance between its nuclear and conventional capabilities which will become increasingly difficult to sustain. If 'something had to go' in 1981 when defence spending was increasing, it's highly likely that conventional forces will be subjected to cuts again in the late 1980s and early 1990s when many new projects will be competing for substantial resources from a reduced defence budget. This may happen in an incremental process with projects stretched, numbers reduced, and ammunition and spare parts cut back. Or it may happen as a result of a more thorough-going review, like that of 1981. Whichever way is chosen, the traditional task of striking the right balance between nuclear and conventional capabilities is likely to become more and more difficult to achieve.

5 Alliance Commitments Versus National Independence

Associated with each of the areas we have so far considered there is a further common problem for defence planners: how to balance alliance membership with national independence of action. Participation in bilateral or multilateral alliances is usually undertaken to supplement a nation's national security interests. Throughout the centuries those states facing common threats have found it advantageous to pool their resources and reinforce their individual power through membership of a wider grouping of states.[1] Alliances, however, can also create obligations and commitments which at times may restrict a nation's freedom of manoeuvre. On occasions there can be, and often are, conflicts of interest between the responsibilities of an alliance and the pursuit of purely national objectives.

Throughout the post-war period one of the crucial tasks facing those responsible for defence decision-making in Britain has been to try to secure the benefits of alliance membership, while at the same time preserving as wide a range of independence of action as possible within the defence field. It has been a task which has not been easy to achieve.

Towards the end of the Second World War, it became increasingly clear to British officials responsible for future planning that Britain could not hope to cope alone with the spectrum of security difficulties that she would face in the turmoil of the post-war period.[2] Further consideration was therefore given from 1944 onwards to bilateral and regional alliance concepts designed to enhance British power and reinforce British national interests.[3] The growing recognition that the Soviet Union would emerge as the greatest continental land power after the war led foreign policy-makers and defence planners to consider how best to pursue Britain's traditional balance of power interests. If Britain could no longer achieve her security interests unaided more attention would have to be given to peacetime alliances.

This wartime planning provided the foundations for the post-war policy of the Labour government to establish a West European Security bloc. As Chapter 1 has shown the Dunkirk Treaty and the Brussels Pact helped to consolidate Western European defence as part of Bevin's wider strategy of sustaining close ties with the United States. The special wartime relationship had been crucial in securing Britain's survival against the Nazi menace and those responsible for security were determined that this relationship, above all others, must be continued into the post-war period. Only the United States had the power to match that of the Soviet Union. As a result NATO, SEATO and CENTO were perceived as useful frameworks to help cement Anglo-American cooperation in areas where British interests were seen to be threatened.

For Britain there is little doubt that these alliances provided important diplomatic and military advantages.[4] While there was a recognition that Britain's power had declined, she could continue playing the role of a Great Power in close alliance with the United States in order to protect the wide range of global interests she still retained. Besides the benefits conferred by alliances, there were also restraints and obligations which had to be borne. In particular, close ties with the United States, membership of NATO's military integrated structure and the commitments accepted under the Paris Agreements of 1954 meant that Britain's freedom of manoeuvre in the defence field was increasingly constrained in the 1950s.

When the Korean War broke out in 1950 Britain's vital interests were not directly at stake. Indeed, Britain, unlike the United States, did not perceive the Korean war to be an example of an international communist conspiracy orchestrated by Moscow and Peking. Nevertheless, Britain did have a vital interest in her alliance with the United States. No one in the British Cabinet was taking the American connection for granted at this time.[5] If the United States was to be persuaded to commit its manpower to the defence of Western Europe, then it was felt that Britain would have to fight alongside her ally for a cause which Washington perceived to be vital to the defence of the West. Even though the rearmament programme introduced by the Attlee government brought Britain to the verge of bankruptcy, the price was considered to be worth paying to keep the alliance together.

This concern over the need to maintain both the American commitment to Europe and the bilateral 'special relationship' has been a feature of British foreign policy throughout the post-war period. For policy-makers the benefits of the close ties with the United States have invariably outweighed the disadvantages. Nevertheless, there can be no denying that there have been disadvantages, especially in terms of constraints on Britain's freedom of action. The Suez crisis of 1956 was a very vivid illustration of the limits of the Anglo-American alliance which had a profound impact on the future of British foreign and defence policy. As far as the Eden government was concerned vital British interests were affected by Nasser's nationalisation of the Suez Canal. The attempt to resolve the crisis by force resulted in the most serious confrontation in Anglo-American relations in the post-war period. Faced with a direct clash of interests, the Eisenhower administration went out of its way to undermine the British operation through financial pressures.[6] As a result, Britain was forced to withdraw in the most humiliating of circumstances. Despite the importance of the military and political partnership between the two countries, when Britain tried to use its armed forces to pursue what were seen as important national objectives, the alliance proved to be of little value. Indeed it was seen to be positively counter-productive.

The lessons which Britain drew from the Suez crisis were quite different from those of France which had also suffered humiliation at American hands. For France, Suez reinforced a key lesson of the past. Useful as alliances were they could not always be relied upon to preserve a nation's security. For De Gaulle, who came to power again in 1958, it was vital that France should enhance its independence of decision-making in the field of security. As a result increased emphasis was given to the Force Nucléaire Stratégique (FNS) and to detaching France from the military structure of NATO. For Britain the lessons were somewhat more ambiguous. It was clear to British policy-makers that Suez had demonstrated the limits of alliances and that as a result independent military capabilities were necessary should alliances let Britain down in the future. Great emphasis was therefore placed in the Sandys White Paper on the further development of the British independent nuclear deterrent as the ultimate symbol of British independence.[7] At the same time, however, it was recognised that Suez had demonstrated the difficulties Britain could face without American assistance. The cornerstone of the Macmillan government's foreign and defence policy after Suez was therefore to try to re-establish close ties with the United States. Indeed, no efforts were spared in trying to make the Anglo-American alliance even stronger than it had been in the past. In the defence field this involved a wide range of military agreements designed to tie the defence establishment of both countries more closely together. Clearly, what the Macmillan government was attempting to do was to balance the benefits of the American alliance with a continuing capability to act independently if the demand arose.

This delicate balancing act was particularly evident in the nuclear field. Both the Thor and Polaris Agreements of 1958 and 1962, respectively, are most interesting in this respect. Under the Bermuda Agreement of March 1957 Britain was to receive sixty intermediate range missiles (IRBMs) from the United States.[8] This provided Britain with the military benefit of a rocket deterrent before she could produce one herself. It also provided the political benefit of signalling the restoration of close Anglo-American ties in the defence field after the Suez débâcle. Significantly, the missiles were to be controlled under a dual-key agreement. Britain would own the missiles and the United States the warheads. This meant that while the missiles helped to reinforce the nuclear forces of the Western alliance as a whole, Britain retained a veto over their use. British sovereignty and independence were preserved. At the same time, however, the United States could veto British use of the weapons, thus undermining Britain's independence.

The Polaris Agreement reached at Nassau, in December 1962, also reflected the British government's search for a balance between independence and interdependence in the nuclear field. On American insistence the Polaris missiles sold to Britain were to be assigned to NATO. They were to be a part of an alliance nuclear force controlled and targeted by SACEUR. The agreement also allowed for the removal of the British missiles from alliance control if, and when, Britain's vital national interests were threatened. But when else would a British government consider the use of nuclear weapons? Since the late 1960s, when the Polaris force was deployed, Britain's nuclear deterrent has been designed to promote both alliance and national interests. Ultimately, though, the determination to retain national independence has taken precedence.[9]

The situation, however, is more complicated than this. Although Britain retains the right, and probably the ability, to fire its strategic nuclear weapons independently, that independence of action is itself somewhat constrained as a result of the alliance relationship with the United States. In the continuing search for greater independence in the post-Suez era through the possession of an independent nuclear capability, Britain has increasingly faced the problem of becoming more and more dependent on the United States . The V-Bomber force and the proposed Blue Streak missile were essentially British delivery systems using British-made nuclear weapons. With the 1958 Atomic

Energy Agreements with the United States, the 1960 Skybolt Agreement, the 1962 Polaris Agreement and the more recent Trident Agreements of 1980 and 1982, a process of interdependence has occurred which in some important respects has left Britain dependent on the United States in a technical and political sense.[10] The continuing requirements for materials, intelligence, information, facilities and components may not prevent Britain from launching its missiles in the short term (except perhaps against some specific counter-force targets), but in the longer term a breakdown of the Anglo-American nuclear relationship would lead to an increasing degradation of the force.[11] Even if Britain was technically able to do without the United States – which is doubtful – it would be extremely expensive and it would take a long time to replace all the benefits of the transatlantic nuclear partnership. The recognition of this fact means that Britain is 'forced' to keep its general strategic and diplomatic policies largely in line with those of the United States.[12] As a result Britain's freedom of action has been, and remains, somewhat constrained.

This does not mean that Britain is not critical of the United States or that it does not pursue its own independent policies. Both Wilson and Heath pursued a number of policies which were largely independent of the United States. The 1980s have also witnessed a number of disagreements over a variety of things from American interest rates to the Yamal pipeline controversy. Britain also undertook a major military action to recover the Falkland Islands in 1982 for purely national rather than alliance reasons. Despite all this, there were limits to how far Wilson and Heath were prepared to cut themselves off from Washington, and there can be no doubt that the government of Mrs Thatcher has been one of the most supportive, if not *the* most supportive, of America's major allies in recent years. On major issues of foreign and defence policy the British government has invariably adopted a pro-American stance, indicating the continuing importance attached to the Anglo-American alliance and the American role within the Western alliance. When the United States has sought diplomatic and military support in recent years Britain has usually been willing to provide what assistance it could. This was the case in the Lebanon, in 1983–4, with the multinational force.[13] Similarly in 1986, despite domestic criticisms, Mrs Thatcher allowed the United States to use American bases in Britain to bomb Libya in retaliation for Colonel Gaddafi's support for international terrorism.[14] Britain was also the first of America's allies to support American naval patrols in the Gulf region, in the latter half of 1987, against Iranian attacks on commercial shipping. Every effort, however, was made to give the impression that British actions were designed to achieve specifically British interests. Coordination of British and American policies was played down.[15] In each of these cases tensions existed between alliance interests and Britain's concern to retain her sovereign independence of action. Despite powerful reservations, in each case the government decided that Britain's interests were best served through support for her American ally.[16]

The question of dependence, interdependence and independence can also be seen in another area of British nuclear policy: that of nuclear targeting doctrine.[17] During the period from the end of the Second World War until the mid-1950s, British strategic policy, as we have seen, increasingly came to be based on deterrence through the threat of a nuclear offensive against an aggressor. Until the British nuclear force became operational towards the end of 1956 this could only be delivered by the American Strategic Air Command (SAC). While the United States had a limited number of nuclear weapons it was inevitable that the so-called 'Sunday punch' would be delivered against targets which specifically reflected American interests. In this respect, American nuclear weapons were targeted primarily against Soviet industry and population centres.[18] Meanwhile, the defence planners in Britain viewed their prospective nuclear capability as an adjunct to that of the United States rather than as a force which would allow Britain to stand alone. It was clear that Britain would need to be able to attack those targets which were not of direct interest to the United States. As Winston Churchill commented in 1955:

> Unless we can make a contribution of our own ... we cannot be sure that in an emergency the resources of other powers would be planned exactly as we would wish, or that the targets which would threaten us most would be given what we consider the necessary priority in the first few hours. These targets might be of such cardinal importance that it could really be a matter of life and death for us.[19]

Britain had an especial interest in an ability to attack Soviet submarine bases, as well as medium-range bomber forces which posed a direct threat from 1954 onwards.

Despite these distinctly British concerns there were two main problems. The first, and most obvious, was that Britain did not possess the ability to pose a serious threat to the Soviet Union until the end of 1956 when the first V-Bomber Squadron became operational. Until that time, therefore, Britain did not have the capability to deal with targets which might be neglected by the United States but which were of crucial importance to British security interests. Secondly, there was also the problem that until 1952 British strategic planning had to take place in something of a vacuum because the United States refused to give Britain any detailed information about American targeting plans. It was not until January 1952 that Churchill was given a personal briefing on the US Strategic Air Plan by the Secretary of State, Dean Acheson. By the end of the year the British Chiefs of Staff had also received their own briefing on a 'highly personal basis'. Even then, although they were kept well informed, 'they were unable to communicate any of this to British commanders in chief for planning purposes'.[20]

The problem of trying to coordinate nuclear strategy with the United States without any detailed knowledge of American plans (until the early 1950s) was compounded, as far as Britain was concerned, by the fact that the United States had been allowed to establish air bases in Britain from 1948 onwards which could be used for a nuclear attack on the Soviet Union in the event of war. As the government increasingly became more aware in the late 1940s and early 1950s that Britain would be a major target for the Soviet Union in the event of war, so the anxiety to know about American strategic planning grew stronger. Matters were made worse by the fact that the 'gentleman's agreement' which set up the American bases in Britain was extremely vague. Britain had no automatic right of consultation with the United States before the President decided to initiate the use of nuclear weapons, and no treaty-based right of veto over the use of American nuclear forces from British territory. Even when the government attempted to establish its sovereign right of veto in late 1950 and 1951, the so-called 'Truman–Attlee Understandings' which resulted, only allowed for 'joint decisions by HM Government and the US Government in the light of the circumstances prevailing at the time'.[21] Although American officials accepted Britain's right of veto in discussions with the British it is these rather vague 'understandings', confirmed during Churchill's visit to Washington in January 1952, which have remained the basis for the American military presence in Britain ever since. It appears, therefore, that once again, in the interests of the alliance, Britain has been prepared to rely on a degree of trust in her relations with the United States rather than demand a formal agreement clearly establishing her sovereignty and independence in this vital area.

While Britain has been content to accept some ambiguity over the use of American air bases in Britain she has been more determined to establish an independent nuclear force from the late 1950s onwards. With the arrival of the V-Bomber force and a thermonuclear capability from 1957, Britain's earlier dependence on the United States was ended. Randolph Churchill told the American Chamber of Commerce in London in November 1958 that, 'Britain can knock down twelve cities in the region of Stalingrad and Moscow from bases in Britain and another dozen in the Crimea from bases in Cyprus. We did not have that power at the time of Suez. We are a great power again.'[22] Thus it appears that by the late 1950s Britain had regained some of her freedom of manoeuvre and possessed an independent nuclear capability which could be targeted on the basis of largely national strategic objectives if so desired. Randolph Churchill's 1958 statement suggested that these targets were Soviet cities.

At the same time, the 1958 Defence White Paper argued that NATO strategy was based on an ability to pose a threat to 'the sources of power' in the Soviet Union.[23] As Lawrence Freedman has noted the phrase 'sources of power' suggested a more 'discriminating range of targets' than just cities.[24] If Britain were to take part in a joint attack with the United States, would the V-Bomber force be assigned different targets from those that would make sense if Britain was acting alone? Freedman identifies the growing ambivalence in British nuclear targeting policy in the late 1950s when he argues: 'Here we have emerging the familiar tension in British nuclear policy: Was the force to be a contribution to NATO's overall deterrent, in which case it had to fit in with the large schemes of the United States, or could it serve by itself as a "formidable deterrent".'[25] A national nuclear deterrent demanded a force capable of destroying Soviet cities. As part of an alliance nuclear force British nuclear weapons would be required to destroy a wide range of targets, including Soviet military capabilities and facilities. The greater accuracy required for counter-force targeting suggested the need for a different kind of nuclear capability from that required to destroy counter-value targets.

In the late 1950s and early 1960s this tension was probably not too acute. At its peak the V-Bomber force consisted of 159 aircraft which together with the sixty Thor missiles based in Britain (until 1963) would have been capable of threatening over 200 targets in the Soviet

Union. Given the size of the force it could be included in the American Single Integrated Operations Plan (SIOP) from December 1960 and in a purely British targeting plan. By the late 1960s, however, the situation had changed. The V-Bombers had been transferred to a largely tactical role and the Polaris force which replaced it in the strategic nuclear role was able to cover only a fraction of the targets covered by the manned-bomber force. If only one submarine could be guaranteed to be on patrol at any one time, then planning had to be based on the possibility that the Polaris force might only be able to attack sixteen rather than over 200 targets.

The increasing tensions in British nuclear policy were reflected in the ambiguity of the Nassau Agreement itself. Under the arrangements worked out with the United States, the Polaris force and other 'allocations' from Bomber Command were to be 'assigned as part of a NATO nuclear force and targeted in accordance with NATO plans'. The British nuclear force henceforth was to be part of the Nuclear Operation Plan (NOP) under the control of SACEUR.[26] The focus of the NOP was on military 'retardation' targets.[27] The problem for British and NATO planners has been that the relatively inflexible Polaris force is not particularly good for attacks on specific theatre military targets. It is a much better counter-city weapon system. As Chapter 4 has pointed out the Nassau Agreement also accepted that, although the force should be assigned to NATO, in the last resort Britain should retain national control over the missiles. The Polaris force has thus been targeted by both Britain and the alliance (in separate plans) against a range of different targets, despite the fact that its capabilities make it a very much better 'ultimate guarantor of national security' than a contributor to alliance theatre military strategy.

The 'Chevaline' programme (which came to fruition in 1982) did little to improve the flexibility of nuclear targeting. The improvement of the Polaris force's capability to reach its targets was begun in the early 1970s when the Soviet ABM programme threatened to make it increasingly difficult to hit Moscow in the future. 'Chevaline' provided a much better chance of penetration in the event of a Soviet deployment of an effective Anti-Ballistic Missile defence around Moscow. Despite this improved penetration, Britain still only possessed the capability to attack 'a few and possibly no more than one large, target(s)'.[28] This doesn't mean that Britain can only attack Moscow but it does rule out the ability of hitting many counter-force targets. Neither does it mean that the Polaris force is a redundant part of alliance plans. It could be used against large military targets, as Alford has pointed out, like dockyards and airfields.[29] The government's frequent description of Polaris as 'a last resort' weapon system, however, does suggest that in terms of the balance of priorities the national rather than the alliance tasks of the force must take precedence. Given the importance of keeping the force in reserve for the ultimate purpose of deterring an attack on Britain itself (and given the limited capability available), it seems unlikely that the government would allow the force to be 'squandered' in the early stages of a conflict by alliance commanders.

The Trident Agreements of the 1980s promise to introduce a new element into the balancing act between the alliance and national purposes of Britain's nuclear force. The Trident system will vastly increase Britain's flexibility in nuclear targeting when it comes into operation in the mid-1990s. Each D-5 missile will have eight MIRVed warheads and, given the extended periods between refits, two and sometimes three submarines will be on station at any given time. This would give Britain at least 384 warheads available compared with the sixteen now available with only one Polaris boat guaranteed on station. As a result of this change, the government undertook a review of targeting policy in the early 1980s in which the greater flexibility in contingency planning was considered.[30] In line with normal practice there has been no public announcement about the result of this review. Nevertheless, the government has made it clear that deterrence must be based on a capability to pose serious threats to 'key aspects of Soviet state power'.[31] Although the precise meaning of the phrase 'key aspects of Soviet state power' has never been spelled out, government spokesmen have indicated that the ability to hit a range of targets other than Soviet cities would be useful.[32]

The greater targeting flexibility which Trident will bring, however, raises renewed questions relating to the priority to be given to national and alliance options. Lawrence Freedman has argued that the targeting debates of the 1980s have produced a significant dilemma for the government. According to Freedman:

> All this leads, inevitably, to the vexed question of the relationship between national nuclear plans and those of NATO and whether or not Britain is presumed to be 'standing alone' or acting in concert with the United States. The conclusion within government is not known, except that it was in the category of 'it's very difficult'. A NATO attack would not be initially directed at the sources of

Soviet state power. SACEUR's objectives would be less ultimate. It is possible to see how Britain's nuclear forces, other than its SSBN's, would fit in with SACEUR's plans, but the implications for Britain of a commitment at this level of escalation do not appear to have been thought through. At any rate, despite the assignation of Britain's nuclear forces to NATO, the assumptions and dominant plans surrounding their targeting do not, as far as can be gathered from the public record, naturally fit in with any NATO plans.[33]

The debate about nuclear targeting in the 1980s has been part of a wider debate about Britain's role in alliance defence which has gathered pace since John Nott's review in 1981. For many observers the decisions taken by the review to cut the surface fleet and retain the commitment to BAOR were a clear demonstration of the government's preoccupation with alliance interests at the expense of British national interests. With the Falklands War in 1982 the critics believed they had been proved right. According to this argument, if the Nott cuts had been implemented by the time the war occurred Britain would not have been capable of undertaking the independent military action which had proved so successful. The lessons of the conflict therefore indicated, so it was argued, the need for more military flexibility to undertake national operations, even if this occurred at the expense of alliance commitments.[34]

Lord Hill-Norton (a former CDS and Chairman of NATO's Military Committee), writing shortly after the Falklands conflict, argued that Britain needed 'the ability to initiate, control and sustain coercive actions whose outcome will be the preservation of its vital interests'.[35] He accepted that British security still depended on membership of 'a superpower alliance' but this didn't mean 'simply falling-in with a more powerful partner'.[36] Nor did it mean providing forces which were merely contributory in nature. 'It demanded,' he said, 'attributes of independence in the military as well as the political field'.[37] While NATO remained fundamental to the UK's security, the country's contribution to this or any other alliance had to be dictated first and foremost by national interests. In Hill-Norton's words:

Since the Brussels and North Atlantic treaties were signed, and more particularly since the middle 1960s, Britain has been at extreme pains to prove herself the best of European and Atlantic Allies. She has done her national interests considerable damage in the process ... It is now time to re-establish our autonomy as a medium power, to provide forces – as do our Allies – which most effectively safeguard the national interest, and then assign them and use them in Alliance interests wherever possible.[38]

Lord Hill-Norton's advice to the British government was to base its defence and foreign policies on Palmerston's famous dictum 'Britain has no permanent Allies, only permanent interests'.[39]

This neo-Gaullist approach has surfaced a number of times in debates about defence policy since 1945. Not surprisingly, it was especially popular in the Conservative Party after the Suez débâcle. It has become even more prominent in the 1980s as the debates about priorities have become increasingly fierce. Writers such as James Cable, Enoch Powell, Admiral Leach, Alan Clark, Michael Chichester and John Wilkinson have all echoed Lord Hill-Norton's arguments in favour of the primacy of maritime power in the pursuit of a more independent defence policy.[40] Like Hill-Norton they do not advocate neutralism nor do they urge Britain to leave the military integrated structure of NATO (as De Gaulle did in his pursuit of a more independent policy). Neither do they see an inevitable conflict between commitments to allies and the pursuit of independent action. They do, however, argue that the government at present has got the balance between alliance and national interests wrong. Instead of the government bending over backwards to placate her allies (which they argue is happening at present), absolute priority ought to be given to national interests.

These are arguments which the government has undoubtedly taken increasingly seriously since the Falklands conflict. The dilemma nevertheless remains. While recognising the importance of flexible military capabilities in order to reinforce Britain's interests outside Europe and most unforeseen contingencies, the government has also continued to emphasise the need to sustain vital alliance obligations.[41] The Thatcher government has modified in some important respects the structure of the defence programme laid down by John Nott in 1981.[42] At the same time, sensitivity to alliance politics has remained an important element in defence planning since the reassessment of British defence policy which occurred after the Falklands War. The unwillingness to enter into a full-blown defence review, with all the implications for alliance and national commitments and capabilities which would result, is a continuing indication of the government's determination to balance these two strands of policy. Its ability to do this, however, (as well as to balance the other strands of policy already discussed) will depend to a large extent on how it deals with a range of economic, technological and political challenges. It is to these challenges that we now turn.

6 The Economic Challenge

While strongly insisting on the great advantages that are certain
to result from the maintenance of peace, and the reductions of
military and naval expenditure, it is quite as essential to assure
that so long as present conditions last, a well-organized and
effective system of defence is a necessary art of state expenditure
... To maintain a due balance between the excessive demands of
alarmists and military officials, and the undue reductions in
outlay sought by advocates of economy, is one of the difficult
tasks of the statesman.

C. F. Bastable, 1895[1]

These words by a nineteenth-century student of public finance are just
as applicable to the period since the Second World War as they were
to the period he was writing about. All post-war governments have
faced the problem of balancing the need to sustain a strong defence
posture with the essential requirement of trying to maintain a healthy
economy (upon which effective military capabilities ultimately
depend). The continuing economic problems faced by Britain for
much of the period since 1945 have posed a major challenge for
defence planners and are unlikely to become significantly easier in the
years ahead. To understand the nature of this challenge we need to
consider the relationship between defence policy and the economy
which has developed since the Second World War.

Britain's traditional strategic posture has undergone a profound
change during the twentieth century. During the first half of the
century defence expenditure rarely rose above 3 per cent of the Gross
National Product (GNP).[2] Britain possessed an army of around
200 000 men which, together with colonial troops, was designed for
'policing' and 'brushfire' operations overseas. It also provided the
'expandable nucleus' for operations on the Continent should circumstances demand. An efficient Navy of around 100 000 could also be
maintained with moderate capital expenditures, largely as a result of
the slow pace of technological developments during this period. As
William P. Snyder has pointed out, the overseas base system and
control of strategic points on the major trade routes meant that 'the

navy was ... able to protect army deployments abroad and to ensure the overseas trade on which Britain's economic life depended'.[3] As a result of this combination of geography, an overseas empire that furnished cheap manpower and strategic bases, and a favourable technological position, Britain was able to maintain an inexpensive but effective national security position.

The situation since 1939, however, has been very different. During the Second World War the defence effort was clearly an over-riding priority. Over 50 per cent of the Gross National Product was spent on defence and more than 13 million people (i.e. 55 per cent of the working population) were engaged either in the armed forces (5.2 million) or 'defence employment' of some kind. Massive debts amounting to between £3000 and £4000 million were also incurred in the fight for survival.[4]

With this scale of wartime expenditure on defence and external disinvestment it was not surprising that in the immediate post-war period attention switched to the tasks of recovery and the rehabilitation of the civilian economy. It was hoped that the three Services could contract to roughly pre-war levels and continue to perform the traditional tasks of home and imperial defence. Defence spending could then be reduced to the 'normal' peacetime level of around 3 per cent of GNP. These expectations, however, were not to be fulfilled. Demobilisation did occur and attention was shifted to domestic production of consumer goods and the provision of social services. Britain's strategic situation, however, had altered in the post-war period making defence a more important priority. Occupation duties, the onset of the Cold War with the Soviet Union, and colonial responsibilities meant that contraction and demobilisation could only go so far. The development of nuclear weapons also had an important effect. The traditional notion of the 'expandable nucleus' was increasingly invalidated as a planning concept. Larger and therefore more expensive 'forces in being' were now regarded as being more important for defence. It was also felt that, whatever the cost, a Great Power like Britain needed its own nuclear weapons for reasons of deterrence and defence. As a result of these perceived requirements in 1948, there were still one and half million men in the armed forces and although defence spending had fallen dramatically from wartime levels it still stood at 7 per cent of GNP – more than twice the pre-war level.[5]

Even this level of defence was not sufficient to confer a sense of security. The perception of threat and the range of obligations meant that much more could easily have been spent on defence. Nevertheless, acute economic difficulties between 1947 and 1949, together with other national priorities, provided limitations on higher defence spending. As David Greenwood has argued:

> In setting the basic contours of the United Kingdom's defence policy in the immediate aftermath of the Second World War it is clear that decision-makers were influenced by 'a generalized perception' that, over the medium-to-long term future, economic capacity and the likely strength of other claims on resources (would) be insufficient to sustain 'costly' postures; and that, in effect, they practised constraint avoidance.[6]

It was these economic constraints which encouraged the government during this period to continue the 'special relationship' with the United States and the creation of an effective alliance framework for the defence of Western Europe.

The government's attempt to balance effective defence with a healthy economy was severely threatened in June 1950 when North Korean forces invaded South Korea. The crisis produced what Richard Rosecrance has described as 'a magnificent governmental and public response'.[7] Despite the government's anxieties, balance of payments problems and the general state of the British economy, a major rearmament programme was undertaken. Defence spending became the 'first objective' of economic policy as the government raised taxes and cut civil spending.

For a year the programme was sustained. The effects on the British economy, however, were very severe. Chronic capacity problems were aggravated.[8] In particular, engineering and allied industries, already suffering from a lack of machine tools, were put under even greater pressure by the rearmament effort. By 1952 it was clear to the new Conservative government that 'the programme could not be pushed through without seriously upsetting the balance of the economy'.[9] As a result, procurement plans were stretched over a longer period in what became known as 'the long haul'. Even with this modification of the rearmament programme, however, the Churchill government was determined to ensure that there would be no downgrading of defence in national priorities. The adjustment, it was argued, was designed to create a healthy economy in order to continue with the defence effort. In the government's view 'any further substantial diversion ... from civil to military production would gravely impair our economic foundations and, with them, our ability to continue with ... the programme'.[10] Despite 'the long haul', expenditure on arms continued to rise and for a time defence spending continued to absorb over 9 per cent of Britain's GNP.

As the Korean War moved towards a conclusion, and the threats to international security appeared to decline, budgetary constraints began to have an increased effect on Britain's defence effort. The Global Strategy Paper of 1952 and the 'Radical Review' of defence policy between 1953 and 1954 indicated the possibility of doctrinal changes which could bring economies in the defence field. A greater emphasis on nuclear weapons seemed to promise 'defence on the cheap'. Sir John Slessor's idea that 'the dog we keep to take care of the cat can also take care of the kittens' appeared to offer the possibility of cuts in expensive conventional forces which hitherto had been designated for European and overseas roles.[11] Plans to cut conscription in 1955, which had to be shelved with the Suez crisis, were resurrected again in 1957–8 as the Defence Minister, Duncan Sandys, attempted to implement the logic of 'the New Look' with its greater emphasis on nuclear weapons.

The task of achieving an effective defence posture at costs which would not destabilise the economy was not so easy to achieve. The commitments undertaken with the Paris Agreements of 1954, and the ongoing responsibilities for overseas defence throughout the fifties and early sixties, resulted in an underlying tension between doctrine and practice. Despite the focus on nuclear deterrence, circumstances demanded a variety of conventional capabilities. Besides the commitment to the V-Bomber force and later Polaris, the requirement for effective land and air forces in Europe, modern naval forces in the Eastern Atlantic and east of Suez, and airlift for overseas intervention, meant that a high level of defence spending was necessary to sustain such a 'balanced force' posture. A continuing expenditure of over 6 per cent of the GNP on defence, however, when Britain was experiencing prolonged economic difficulties during the 1960s, eventually proved to be unsustainable.

The tendency of governments throughout the 1950s and early 1960s systematically to underestimate attainable growth rates came to a head in the period from 1964 to 1968 with the three major revisions of public spending targets 'prompted by shattered growth illusions'.[12] The task of moderating inflationary demands and restoring balance of payments equilibrium pushed the government towards major revisions in defence policy. The withdrawal from east of Suez, strictly speaking, may not have been 'forced' on the Labour government but the need to stabilise the domestic economy meant that reductions in defence, as well as in other areas of public expenditure, were perceived to be necessary.

Similar economic constraints had their impact on British defence policy during the 1970s and early 1980s. Defence plans laid down by the Heath government between 1970 and 1972 were 'rudely slashed in the interests of macroeconomic management'.[13] Continuing economic difficulties also lay behind 'the most thorough peactime review ever undertaken' in 1974–5.[14] On this occasion the Wilson government planned to save £4500 million over a ten-year period and reduce the burden of defence over the same period to 4.5 per cent of the GNP (from 5.5 per cent). Further cuts in defence were subsequently made in 1975 and 1976 as economic difficulties persisted.

By the mid-1970s, however, evidence was mounting that the pursuit of economic stabilisation was having a serious effect on Britain's defence effort. Public criticisms were voiced by the Chiefs of Staff, the Conservative opposition, and even by American generals, suggesting that the neglect of the nation's armed forces was badly affecting both morale and capabilities. The lack of ammunition and fuel in particular meant that the forces were finding it increasingly difficult even to train properly. Commentators pointed out that while it was important to have a strong economy to maintain effective defence, a serious erosion of the defence effort as a result of inadequate provision could undermine the sense of security upon which the pursuit of economic well-being ultimately depended.[15] The criticisms appear to have had their effect. In an effort to redress the balance the Callaghan government agreed to comply with the 1977 NATO goal of a 3 per cent increase in real terms, per annum, in defence spending.

In the longer term even this proved to be inadequate. Even with a significant increase in defence resources in the late 1970s and early 1980s, the new Conservative government of Mrs Thatcher discovered that it was impossible to fulfil the defence programme laid down by their predecessors. The task of economic reconstruction, with its emphasis on restraining public expenditure, meant that increasing the defence budget even further was unacceptable. If resources could not be increased the only alternative for the Defence Secretary, John Nott, was to cut capabilities. In the words of the 1981 review, 'something had to go'.[16] Once again the search for economic stabilisation and economic growth dictated the need to prune defence (in this case naval) capabilities.

The lesson of the post-war period is that economic problems have continuously prescribed the limits of policy choice as far as British governments have been concerned. David Greenwood is surely right to emphasise that budgetary constraints have not been 'decisive' in shaping British defence policy 'in a strict, imperative sense'.[17] Defence policy has been essentially the product of political choice. Nevertheless, as the analysis of the relationship between the state of the economy and defence policy above suggests, the budgetary constraints are 'a catalyst for choice'. As such they have been instrumental in 'forcing' governments to reconsider their priorities continuously over the post-war period. Whether their influence has been constructive or essentially malign is a matter of dispute, but they have helped in the process of reshaping the nation's defence posture which has undoubtedly occurred during the period since 1945.[18]

The continuing challenge to policy-makers to balance economic growth with an effective defence effort is as difficult to meet at the present time as it has been in the past. Indeed in some important respects the contemporary challenge is even more difficult to resolve. This can be shown by looking at the projected defence spending and the plans for defence capabilities over the next few years.

During the first half of the 1980s, despite the 1981 defence review, the government gave a high priority to defence expenditure. From 1979 to 1986 defence spending rose by more than 21 per cent in real terms, while total public expenditure increased by only 10 per cent in real terms. In spite of the 1975 projection that defence spending was to fall to around 4.5 per cent of GNP in the 1980s, the commitment to the 3 per cent real increase per annum from the late 1970s onwards meant that defence spending settled around 5.2 per cent of GNP from 1981 to 1986.

The decision to end the 3 per cent real increase in defence spending from 1986 has changed the situation considerably. According to the 1986 Public Expenditure White Paper, the defence budget was scheduled to increase by only 7 per cent in cash terms in the three-year period up to 1989. In effect, after allowing for inflation, this represented a cut in real terms of 6.4 per cent. George Younger managed to get a 'marginally better deal for defence in November 1988 for the period up to 1991/2. Defence spending, however, is still only scheduled to rise by 1.7 per cent in real terms in 1990/1 and by 1.3 per cent in 1991/2. These increases also depend on an inflation rate of only 3 to 3.5 per cent.[19]

As we have seen, the 1981 defence review calculated that, even on the basis of a continuation of 3 per cent real increases per annum in defence spending, something would have to go. Since then the government has not only cancelled the 3 per cent commitment but it has decided to purchase the more expensive Trident D5 missile from the United States; it has fought the Falklands War and continued to garrison the islands ever since; it has decided to keep a fleet of 'around' fifty surface ships (even though the 1981 review planned to reduce the number to around forty-two); and attempts have been made to improve Britain's intervention capabilities. At the same time, the government has plans for a range of new conventional weapons and equipment for the 1990s. All this has to be achieved on the basis of more or less level funding for defence.

The challenge facing the government is clear. Can it be done? Many experts outside government have their doubts. David Greenwood has pointed to what he describes as 'a funding gap' between what the defence programme on the books would cost if put into effect as planned and what government documents reveal it intends to spend in reality.[20] The significant disparity has led Greenwood to suggest that hard choices will have to be made and the best way forward is a major defence review which will scrutinise each of the four main pillars of British defence policy.[21]

This view, that major cuts in defence will become necessary unless there are increases in defence spending, is shared by many other observers of the defence scene.[22] *The Sunday Times*, for example, argued in January 1986 that 'the arithmetic of future (defence) spending shows that unless there is a radical rethink matters can only get worse'. The paper's editorial accused the government of trying to 'squeeze a quart of commitments into a pint of resourses'.[23] Much the same kind of criticism was made by opposition parties in the 1987 General Election. The Labour Party, in particular, argued that the government's commitment to Trident would lead inevitably to significant cuts in conventional capabilities in the future.

The government's response has been that a major defence review is not necessary. Sir Clive Whitmore, the Permanent Under Secretary (PUS) at the Ministry of Defence, told the House of Commons Defence Committee in 1984: 'We are not in defence review country . . . in the normal way we will have to make "adjustments to the programme" to bring it into line with the defence budget'.[24] According to the Ministry of Defence the 'better value for money' approach could largely solve the problem. Better management techniques, fixed price contracts, more privatisation and the stretching of procurement were seen as the techniques to see the Ministry through its difficulties.[25]

It was perhaps not surprising that the government should publicly declare its opposition to a major review in the run-up to the 1987 election. The Nott review of 1981 had caused the government considerable embarrassment, given its declared commitment to defence in the 1979 election. In particular, the decision to cut the surface fleet had been strenuously opposed by the naval lobby which included a number of Conservative backbenchers. A further review before the 1987 election would have renewed unease amongst government supporters and deflected attention away from the Conservatives' campaign against Labour's defence policy. It would have been difficult for Mrs Thatcher to have accused her political opponents of being soft on defence when her government was engaged in a major exercise designed to prune defence capabilities. The political circumstances were clearly against a major public review prior to June 1987.

One of the key questions for the future, therefore, is whether the Defence Secretary will undertake a major review before the next election in 1991 or 1992. There would appear to be three options open to the government. One would be to adopt an ad hoc process of cancelling selected weapon systems, deferring others, cutting the stocks of ammunition, fuel and spare parts, and generally 'spreading the misery' equally across the board. To a large extent this incremental process of 'review by stealth' has already been under way for some time.[26] The second option would be to go for a thorough-going defence review which focuses on priorities and considers which of the major defence capabilities should be cut in the interests of the total defence effort. And the third option would be significantly to increase the resources devoted to defence so that 'the funding gap' can be closed and the government can implement the full range of its defence plans. All these options, however, have advantages and disadvantages as Chapter 9 will try to show.

Whichever is chosen the fact remains that the government's predicament arises largely as a result of the powerful economic constraints on defence policy. No government has unlimited resources to devote to defence. Striking the right balance between maintaining a healthy

economy and providing effective defence has been a major preoccupation of all British governments since 1945. The evidence suggests that
this is a challenge which is likely to continue to face defence planners
and may well become very acute during the 1990s.[27]

7 The Technological Challenge

Economic and technological challenges have been closely linked in the history of British defence policy since 1945. Important developments in land, sea and air warfare occurred in the first forty years of the twentieth century but these were much less significant than the quantum leaps in military science which have taken place in the period since 1945. As John Garnett has written, 'the sheer pace of technological innovation in weapon technology ... is one of the startling features of our time'.[1] In less than a century mankind has moved from the horse-drawn artillery of the First World War to the precision-guided munitions of the 1980s and the ideas of space-based lasers and particle-beam weapons envisaged in President Reagan's Strategic Defense Initiative. In attempting to adjust to the rapid process of technological change defence planners in Britain, like those in other countries, have been faced with numerous difficulties. In the main there have been problems of cost, centring largely on debates over quality versus quantity; problems of uncertainty in defence planning; problems of dependence and interdependence; and problems of adjustment as tactical and strategic doctrines have evolved.

The problem of cost has been very much at the heart of the challenge posed by technological innovation over the post-war period as a whole. Working at the forefront of any scientific field, pushing the boundaries of knowledge to the limits, is inevitably an expensive process. This has been especially true in the field of military science where the costs of equipment have taken an ever larger slice of the defence budget. In 1950, 30 per cent of the defence resources in Britain were allocated to equipment. In 1975, this had risen to 35 per cent. By 1984 it was 46 per cent and rising. The price of army equipment in the 1960s was four times what it had been twenty years before; the other Services being affected in a similar way. The 1984 Defence White Paper revealed that, 'The new Type 22 frigate is three times as expensive in real terms as the Leander; the Harrier is four times the cost of the Hunter; and a new artillery shell is double the price of its predecessor.'[2] The White Paper admitted that one of the most serious problems facing defence planners was 'how to contain the seemingly inexorable rise in the cost of defence equipment'.[3]

As defence inflation has been on average 6 to 10 per cent above normal rates of inflation, the numbers of units the government has been able to buy (from a defence budget which has remained reasonably constant in real terms) has inevitably declined. By the 1980s Britain had only half the number of planes it had twenty years before. The Navy had only half the number of destroyers and the Army 300 fewer tanks. To be sure, the aircraft, destroyers and tanks of the 1980s are much more sophisticated and much more effective in many ways than those of the 1960s. It is also true that qualitative improvements are essential to meet the growing sophistication of the threat. The 1984 Defence White Paper pointed out:

> There is no sign of any slackening in the pace of technological change. To meet the increasing sophistication of the threat the quality of defence equipment must continue to grow for the foreseeable future. This brings its own benefits in terms of increases in reach and hitting power.[4]

The constant pursuit of quality at the expense of quantity, however, clearly has its limits. The danger of acquiring fewer and fewer items of increasingly sophisticated equipment leads down 'the road of absurdity'. At the end of that road is one incredibly sophisticated ship, plane or tank. However technologically advanced certain items of military equipment are, there are obviously occasions when numbers are much more important. Small numbers of even the most sophisticated equipment can be swamped by sheer mass. Clearly the contraction of defence equipment cannot go beyond the point where force structure and levels no longer allow the Services concerned to perform their specific roles effectively. The government's dilemma is how to balance the need for quality in certain areas where only quality will do, with quantity in other areas where sufficient numbers are essential. With technology constantly changing, it is never an easy balance to achieve.

Problems of cost also arise because of the uncertainty which rapid technological change brings. Given the long lead-times of modern weapon systems of between ten and twenty years, it is extremely difficult to plan in terms of the threats and kinds of technology which will be available this far ahead. Working at the forefront of technology also means that sometimes weapons do not live up to expectations. In the past this has affected British defence policy in two ways.

First, as the technology or the threat has changed certain weapons have had to be cancelled before they have been completed. Between 1952 and 1965, for example, more than £300 million was invested in air projects which were subsequently cancelled.[5] The problems have not got significantly better since then. During the 1980s the same fate overtook Britain's Nimrod early-warning aircraft.[6] Escalating costs and failures to resolve technological difficulties caused the cancellation of the project after millions of pounds had been spent. The government then had to purchase the AWACS system from the United States at a further cost of £860 million.[7]

A second effect has been that, as manufacturers have attempted to keep up with the threat and the technological advances which are continuously being made, weapons systems have had to be changed and modified during the procurement process itself. Such changes in specifications and capabilities from those originally planned lead to extra costs and sometimes less than satisfactory solutions to problems have to be accepted. During the production of the Tornado GRI constant attempts were made to upgrade the aircraft's radar to allow it to operate effectively in the changing technological environment. Even when the aircraft did eventually become operational in 1981, at a cost of £17 million each, the radar had still not been perfected. Improvements had to be planned for later models.[8]

The problems of uncertainty caused by technological advances are not confined to the procurement process. Defence planners constantly have to face the problem of how to incorporate new and powerful technology into tactical and strategic planning.

This was particularly a problem for Britain at the end of the Second World War. The bombing of Hiroshima and Nagasaki in 1945 ushered in a new and revolutionary atomic age to which defence planners had to adjust. Although the full implications of these startling technological developments were still unclear in the immediate post-war period, there was nevertheless a recognition that the ability of atomic weapons to bring about instantaneous mass destruction meant that a radical change in traditional strategy would be necessary. The focus henceforth would need to shift from how to fight and win wars to how to deter them. Nuclear deterrence, albeit in a crude form, therefore became a central strategic concept as defence officials attempted to formulate a suitable doctrinal framework into which to fit the new technology.

The task of combining the new ideas of nuclear deterrence, which emerged in the late 1940s and early 1950s, with traditional ideas of conventional defence proved to be problematical. In 1952, the Global Strategy Paper was produced by the Chiefs of Staff in an attempt to bring together the conventional and nuclear elements of British strategy into one coherent plan. The Global Strategy Paper recognised the need for some conventional capabilities in Europe and for Cold War operations but the pride of place was given to nuclear weapons. Nuclear deterrence was to hold the centre of the stage. This at least was the theory. In practice a tension soon emerged between strategic doctrine and practice. Circumstances increasingly demanded an important conventional role for each of the Services compared with the limited conventional capabilities envisaged in the Chiefs' paper. As Richard Rosecrance has pointed out:

> Cold war forces were supposed to be increasingly outmoded, and yet they occupied British attention and resources from 1953–7. Kenya, Malaya, and the Middle East were of great importance, regardless of their defensibility in thermonuclear war. The Air Force continued with fighter-interceptor and night and all-weather re-equipment along lines foreshadowed in the Labour rearmament program. The Navy went on dealing with the menace of the submarine and mine and studied the problem of countering Soviet cruiser strength. The Army maintained its European ground forces commitments, though they failed to reach the Lisbon goals projected for 1954. In one sense . . . the global strategy paper, instead of producing cuts in conventional forces to re-emphasize 'the great deterrent' actually sanctioned a strategic rearmament alongside conventional rearmament.[9]

What Britain attempted to do in the early 1950s, therefore, was to superimpose a nuclear strategy on top of a conventional strategy rather than integrate the two elements into a coherent strategic plan. As a result, as John Strachey told the House of Commons in 1955, Britain 'was attempting to get something of everything and succeeded in getting enough of nothing'.[10]

The task of adjusting strategy and tactics to the new nuclear technology was also experienced in the 1950s when smaller tactical nuclear weapons began to appear. These were weapons which had never been used in actual combat conditions and whose performance thus remained problematical for defence planners. One of the key difficulties centred on whether conflicts in which these weapons were used could be kept limited or whether, once the nuclear threshold had been crossed, total war would inevitably result. In the 1950s, the Chiefs of Staff had grave doubts about the notion of limited nuclear war.[11] Nevertheless, Britain produced a range of tactical nuclear capabilities for operations in Europe even though planning on their use was rather vague. These doubts remained in the early 1960s when the United States pressed its NATO allies to accept the doctrine of Flexible Response in which the notion of limited nuclear war and controlled step-by-step escalation figured prominently.[12] At one level British planners recognised that limiting a nuclear war should it break out was attractive (given the alternative). But at the same time they remained sceptical about the possibility of maintaining the limitations once the first tactical nuclear weapon was used. Even in 1967, when Britain (along with its allies) accepted a Flexible Response strategy for NATO, the scepticism about the tactical role of nuclear weapons in the strategy remained. This feeling has continued in certain quarters of the British defence establishment down to the 1980s.[13] Given the nature of the technology and the uncertainty associated with its use this is perhaps not surprising.

The ambivalence over the role of tactical nuclear weapons in British (as well as alliance) strategic and tactical planning may be eased in future by the emerging technologies of the 1980s. Instead of the steady, incremental improvements to conventional weapons we have witnessed for the last forty-odd years, we now appear to be on the verge of a quantum jump in conventional technology. It seems likely that a range of new technologies have the capability of revolutionising the effectiveness of non-nuclear weapons, even to the point of enabling NATO to assign conventional weapons systems to missions previously regarded as exclusively nuclear.[14] If this proves to be the case the traditional unease over the tactical role of nuclear weapons may be significantly eased. Non-nuclear technology may become the force multiplier which nuclear weapons have been since the early 1950s.

At the root of this potential revolution in conventional technology is the incorporation into weapons and reconnaissance systems of advanced data-processing systems and a variety of optical, radar, infra-red and laser sensors that offer extraordinarily accurate target acquisition in all types of climatic conditions and at a great distance. Emerging Technology opens up the opportunity for military commanders to perform their tasks much more effectively than ever before. New reconnaissance and surveillance techniques will make it easier to identify enemy targets. New data-processing and communications technology will improve the ability to pass information to those that need it more quickly and effectively than has been possible in the past. The greater accuracy and lethality of modern weapons, especially precision-guided munitions (PGMs) will improve the ability to destroy selected targets. At the same time, 'stealth' technology may improve the ability of some weapons systems to avoid destruction. New technology therefore seems likely to have an important impact in all areas of modern war.[15]

Besides the benefits, emerging technology brings with it numerous problems which pose important challenges for defence planners, such as the perennial problem of cost, and the difficulty that much of the new technology comes from the United States which threatens the future defence industries in Europe. Some of it has distinctly offensive characteristics which makes it difficult for a defensive alliance like NATO to utilise without political difficulties. And above all, because by its very nature it is *emerging* technology, there is the great uncertainty about its use. As John Garnett has argued, 'E.T. has all sorts of tactical and strategic implications, but since many of the new weapons are not yet developed and few have been tested in conditions of war, no one can predict reliably what the consequences will be'.[16] How exactly British and alliance defence planners will absorb the wide range of new technology into their tactical and strategical doctrines is not clear at present. FOFA provides one example of how the alliance is attempting to build on the foundations of existing concepts. The full implications of the new technology, however, remain uncertain. The danger is that an increased emphasis on offensive technology will erode the basic defensive orientation of the existing strategy.

It will take many years of detailed planning and creative thinking before all the aspects of the conventional revolution of the 1980s are fully translated into operational military concepts. The difficulties of devising and developing strategic and tactical doctrines for a world where weapons change so quickly are far from easy to resolve. The British government recognised the difficulties in the early 1980s when it attempted to reassess the impact of technological changes on the balance of defence capabilities. The June 1981 White Paper, *The Way Forward*, accepted that:

Technological advances are sharply changing the defence environment. The fast-growing power of modern weapons to find targets accurately and hit them hard at long ranges is increasing the vulnerability of major platforms such as aircraft and surface ships. To meet this and indeed exploit it, the balance of our investment between platforms and weapons needs to be altered so as to maximise real combat capability. We need to set for the long term, a new force structure which will reflect in up-to-date terms the most cost-effective ways and the key purposes of our defence effort.[17]

The government recognised, however, that adapting to technological change would not be easy. 'Moving in this direction', the White Paper argued, 'will mean substantial and uncomfortable change in some fields.'[18] The subsequent difficulties which arose in implementing the Nott review demonstrate just how hard it is to break down the rigidity of past patterns of defence effort. This is a challenge which the government will continue to face.

The challenge and uncertainty created by new technology in the 1980s has not been confined to the conventional field. The continuing Soviet research effort into ballistic missile defence, and the more public Strategic Defense Initiative launched by President Reagan in March 1983, has raised question marks in Britain about the future viability of Britain's nuclear deterrent force. The development of nuclear weapons and missile technology in the 1950s and 1960s brought a revolution in contemporary strategy which ushered in an era of Mutual Assured Destruction (MAD). Deterrence was maintained through the vulnerability of both East and West to nuclear devastation. President Reagan's rejection of the moral and security foundations upon which MAD was based and the search for strategic defences has raised important questions in the 1980s about whether new technology will usher in an era of more stable deterrence based on defence rather than the threat of retaliation. Should this prove to be possible the effects would be profound.[19] The movement in strategic thinking away from Mutual Assured Destruction to Mutual Assured Survival would clearly have a significant effect on the British nuclear deterrent. Britain might have spent in the region of £9 billion on the new Trident system which will become operational between 1994 and 1997 only perhaps to find that new technology has made it obsolescent. Strategic defences could undermine the whole retaliatory *raison d'être* on which Trident is based.

Whether or not this will occur remains distinctly uncertain in the late 1980s. Important question marks remain over the technological feasibility of the SDI programme and the determination of future American administrations to pursue the programme through to its completion. What seems certain is that even if the United States gives up the drive for a total leak-proof defensive system, the research effort by both superpowers to discover exactly what is technologically possible will continue. Just as the Soviet ABM programme created difficulties for Britain in the early 1970s and required the search for technological refinements to the Polaris programme, so also the uncertainty created by Star Wars research in the years ahead will pose important and probably expensive challenges to those responsible for Britain's nuclear programme.[20]

Britain's nuclear deterrent force also provides an illustration of another technological challenge which affects every aspect of British defence policy. This is the problem of independence and dependence. It is the ideal of every defence planner to pursue policies which, in the vital field of the security of the realm, will be as independent of outside control as possible. Britain's nuclear programme in the post-war period demonstrates how difficult in reality this is to achieve.

As Chapter 4 has shown, Britain's pursuit of an independent nuclear deterrent began in 1947. Largely unaided by any other state, Britain produced both atomic and thermonuclear weapons during the 1950s and an effective delivery system based on the V-Bomber force. As technological developments increased the vulnerability of the manned bomber, Britain attempted to retain an effective nuclear capability through the development of Bluestreak, a liquid-fuelled intermediate range ballistic missile. Although Britain obtained some assistance from the United States, Bluestreak was largely a project of home-grown technology. The government could still claim, with some justification, that Britain would retain an independent nuclear system into the 1960s. The exacting technological requirements of Bluestreak, particularly the need to place the missile in underground silos, increasingly pushed up the costs in the late 1950s and in 1960 the government reluctantly decided to cancel the project.

The pursuit of true independence had finally ended. With the 1958–9 Atomic Energy Agreements with the United States to collaborate on nuclear research and the 1960 decision to purchase American Skybolt missiles, the age of nuclear interdependence began. To be sure, Britain continued its own nuclear programme and Skybolt was designed to enhance the life of its own V-Bomber force. Nevertheless, in the face of the growing technological challenge the government had given up the task of trying to maintain total independence.

Henceforth, although successive British governments were to insist on retaining the ability to launch nuclear weapons independently, the Skybolt, Polaris and Trident Agreements represented the acceptance by Britain of a substantial degree of dependence on the United States. The pace and cost of technological change may not have been the only considerations which forced Britain towards diluting her independent nuclear capability – after all France had decided to maintain her national nuclear force although faced by similar pressures. If Britain wanted to stay at the very forefront of nuclear developments and retain what the government regarded as an effective nuclear deterrent, there was no choice but to depend on the United States.

Whether the growing technological dependence seriously constrains Britain's foreign and defence policy has been in the past, and remains today, a matter of major political dispute. When Skybolt was cancelled unilaterally by the United States in 1962 the British government found itself in a very awkward position and vulnerable to criticisms from the Labour opposition.[21] Throughout the life of Polaris successive British governments have been forced to defend the nuclear deterrent against criticisms that the range of facilities, materials and components provided by the United States makes a sham of the claims to independence. A similar row arose over the purchase of Trident, especially when the government was forced, in 1982, to switch from the C4 missile which it had originally purchased in 1980 to the D5 missile, simply because the United States had decided to phase out C4 earlier than planned. Revelations in October 1987 that Britain's Trident missiles were to be refurbished in the United States at King's Bay, Georgia, rather than at Coulport in Scotland, where Polaris missiles have been overhauled, led to renewed accusations that dependency was now virtually complete.

From the Labour backbenches Denis Healey told the House of Commons on 29 October 1987 that Britain faced

> a prolonged and humiliating dependence on the US which will cover the whole of our foreign and defence policy ... the 'rent-a-rocket' Trident programme – in which the missiles will be exchanged from the US pool every seven or eight years, and British made warheads had to be tested on a US test-range – made Britain totally dependent on the Americans and totally incapable of standing up to them.[22]

Healey argued that the dependency was corrupting areas of British foreign policy not strictly relevant to the issue. Britain was the only ally to give the United States a facility for the raids on Libya, he said, and 'we were pussyfooting on Central America, and following the US line on the Gulf'.[23] There was, he said, a basic contradiction in British policy brought about by dependency. Britain needed Trident in case the United States let us down in a crisis, but 'if we cannot rely upon them in a crisis can we rely upon them to provide us with Trident?'[24]

The government's reply was that the label, the 'Moss-Bros missiles', was quite unfair and the whole notion that Britain was 'renting' Trident was untrue. The Defence Secretary, Mr Younger, argued that Britain still retained full control over the missiles. The refurbishing agreement, he argued, saved the government £700 million and the arrangement would not, in any significant way infringe the truly independent nature of Britain's Trident fleet. For the government the King's Bay Agreement was part of a long-standing policy of making use of a range of American facilities which would allow Britain to retain a nuclear deterrent as cheaply as possible.[25]

Despite what the government said, acquiring key items of Britain's nuclear deterrent from the United States does provide Washington with some control over the force. Attempting to save money by increasing the use of American facilities inevitably brings greater dependence. The challenge is to make sure that the erosion of independence does not reach the point where the short- and medium-term ability of the British government to launch the missiles is seriously compromised.

The problems of dependency are not, of course, only to be found in the nuclear field. The rapid pace of technological innovation and the concomitant costs create pressures for greater interdependence and cooperation across the whole spectrum of defence capabilities. During the 1970s Britain entered into a wide range of bilateral and multilateral agreements with allies in Europe and across the Atlantic. This involved projects like Jaguar, Tornado, the EH101 and NH90 helicopters, MLRS, and more recently the European Fighter Aircraft (EFA). Many of these projects have proved very successful providing Britain with technologically advanced weapon systems which she would have found it difficult to produce by purely national means. Developing even greater European defence cooperation in the future, in the face of American and Japanese competition, would appear to be of great importance to Britain and her allies.

Cooperation in the field of high technology also has its own problems. Cancellations, divergent objectives and requirements between nations concerned and the complex difficulties associated with multilateral management of advanced weapon systems, have all produced headaches for British defence planners. The need to be seen to be a good European has imposed political restraints on the commercial aspects of policy. Participation in these projects has also created a degree of dependence in certain key areas of military technology. The economic, technological and, at times, political advantages of cooperation are clear, but there is always a problem of deciding on what kind of balance to strike between cooperation and independence in the development of military technology. To compete effectively with the United States in the future much more market integration in the high technology areas is going to be necessary. In addition, retaining a strong national technological base is going to remain important if Britain is itself going to compete in the world market. As the pace of technological change quickens and costs rise, the task of achieving the right balance is going to become more and more difficult.

The need to confront the technological challenge is going to be one of the most vital and fundamental tasks for defence planners in the future. Every aspect of defence is affected by it. The problems of cost, uncertainty, doctrinal and force structure adjustment, dependency and interdependency all contribute to the growing complexity of future policy planning.[26] Above all the sheer pace of technological innovation at the present time creates unprecedented problems. Defence planners have in the past been used to lead-times of between ten and twenty years and a life-span of weapons and equipment of often more than twenty years. Modern technology, in a variety of fields, now has a much more limited life-cycle. Sometimes this is no more than three or four years. The challenge which this poses is a very formidable one indeed.

8 The Political Challenge

Economic constraints and technological innovation have clearly been very influential in helping to determine British defence policy since 1945. As Chapters 6 and 7 have tried to demonstrate they are also likely to pose important challenges for the future. Defence policy, however, is not simply the product of economic and technological pressures. Such pressures help to define the limits of policy choice but ultimately that choice (relating to the size, shape, equipment and deployment of the armed forces) is a political matter. Defence policy is the result of a judgement about priorities which stems from the values of those who hold power. In the case of democratic governments that power resides in the elected government of the day. In Britain, as Harold Laski has argued

> It is the essence of our Parliamentary system to make the responsi-
> bility for ... priority in value one that the Cabinet must take. Any
> alternative would destroy that system because it would ultimately
> place the responsibility ... outside the area where it can be
> controlled by Parliament and the electorate. That road ultimately
> leads to dictatorship ... For it rests upon the assumption that,
> whatever the popular will, it must accept the specialists conclusion
> of what is good for it.[1]

Although it is the government's responsibility 'to choose and, in so choosing to establish national priorities', as Laski also reminds us, governmental choice has to be sensitive to Parliament, popular sentiment and the electorate directly.[2] It is certainly true that governments lead public opinion as well as reacting to it. However, in exercising this leadership they are not unconstrained. Public opinion cannot be led where it is not prepared to go.[3] In the defence field, as in other areas of government policy, party politics and public opinion therefore play their part in the determination of policy.

One of the most significant events in British domestic politics during the 1980s has been the breakdown of the post-war inter-party consensus on defence policy. With limited exceptions defence did not figure prominently in General Elections between 1945 and 1983 and was rarely an issue of major controversy between the main political parties.[4]

Amongst the general public also there was a widespread acceptance of the fairly high levels of expenditure devoted to defence (compared with our European allies) and the major priorities in British defence policy. This consensus included a continuing perception of a threat from the Soviet Union, the need for a 'special relationship' with the United States, the central importance of NATO and European defence, the acceptance of nuclear weapons and the provision of defence resources to maintain British interests outside the European area. The foundations of British defence policy which encompassed these priorities were laid down by the Labour government between 1945 and 1951 and have been largely upheld by all governments up to the present day.

It is true that within the defence community itself debates and controversies about military priorities have been a constant part of the policy-making process. Certainly the series of defence reviews which have punctuated the period from the end of the Second World War through to the 1980s were occasions when major disputes were fought out between the Chiefs of Staff themselves and between the COS and the government of the day. Nevertheless, with the exception of debates about the Korean War in the early 1950s, disputes over nuclear weapons (with the rise of CND) in the late 1950s and early 1960s, and the controversy surrounding Britain's east of Suez policy in the late 1960s, defence was not an issue which held much interest for the general public. Neither was it an issue which figured largely in the contest between the two major political parties. Despite the occasional clashes on specific defence issues, the Labour and Conservative Parties, once in office, tended to pursue the broad outlines of the defence policy inherited from their predecessors. Bipartisanship was a key characteristic of British defence policy.

In the early 1980s, however, major divisions emerged between the government and the opposition parties which made defence an important issue in both the 1983 and 1987 elections. Such a breakdown of inter-party consensus in recent years clearly poses new and important political challenges which are likely to remain with us up to, and beyond, the next General Election in the early 1990s.[5]

The change occurred towards the end of the 1970s. A conjunction of domestic and international events helped create a growing and sustained public awareness of defence matters in a way which has not happened previously. As a result defence began to move towards the centre of the political stage in Britain and throughout Western Europe. Issues such as technological developments associated with enhanced radiation weapons, 'Chevaline', Trident, cruise missiles, Pershing II, MX and SS20 missiles, and political issues involving the deterioration of East–West relations and the growing disillusionment with multilateral arms control negotiations all helped to heighten the level of public concern over defence and security. Britain, like many other European states, experienced the resurgence of a vociferous 'peace movement' which criticised some of the major planks of Britain's post-war defence policy. Increasingly, Britain's independent nuclear deterrent and the decision to replace Polaris with Trident, together with NATO nuclear modernisation plans, became the subject of wider public debate. It was in this environment that a process of polarisation on defence issues gathered momentum. As the Labour Party and to a certain extent the Liberal party began to reflect the public concern over prevailing nuclear policies, the election of Margaret Thatcher in Britain and Ronald Reagan in the United States brought to power two governments passionately committed to strengthening the West's defence, including its nuclear defences.

During its period in office from 1974 to 1979 the Labour government, initially under Harold Wilson and then under James Callaghan, had sustained the major parameters of post-war policy. Despite periodic Labour Party conference resolutions and Manifesto promises to seek the removal of American Polaris bases from Britain, to cut defence expenditure and to phase out Polaris after it came to the end of its useful life, there were few radical changes implemented during Labour's period in office.[6] Indeed, the Callaghan government accepted the 1977 NATO commitment to increase defence spending by 3 per cent in real terms and even initiated a number of studies on a possible replacement for Polaris. As Peter Jones has argued, there were several reasons for this continuity of policy during the 1970s. The most important of these centred on

the slimness of the government's majority and its reliance on other parties for support during the whole of the period; the character of the party leaders who were both in full control of their party and staunchly pro-NATO and convinced multilateralists; the continued perception of detente as the best hope of maintaining low levels of east-west tension and keeping the escalating cost of defence under control; and finally the need to concentrate on the more immediate economic and political crises following the 1973 Middle East war and the massive increase in oil prices.[7]

What the period from 1974 to 1979 did show was a growing divergence on defence policy between the Labour government itself and the party activists. After the party left office in 1979 and James Callaghan was replaced as party leader by Michael Foot, the divisions within the party came to the fore. The party was now led by a committed unilateralist for the first time since 1945. This change at the highest level, together with the 'defection' of the 'Gang of Four' to form the SDP, helped to drive the party further towards a more radical posture on defence. By the time of the 1983 election Labour had committed itself to a non-nuclear defence policy within the 'lifetime of the next Parliament'. The Trident programme was to be cancelled, Polaris was to be included in disarmament talks and all nuclear bases in Britain were to be closed down. Defence spending was also to be reduced to bring it into line with the other major European NATO countries. Although the Manifesto committed a future Labour government to continuing support for NATO, every effort was to be made to shift NATO towards a non-nuclear strategy.[8]

During the 1983 election these radical proposals were opposed vigorously by both the Conservatives and the Liberal/SDP Alliance.[9] The Conservatives claimed to be pursuing a defence policy which had been accepted by all post-war governments and accused Labour of putting Britain's hard-won security at risk because of the 'reckless and naive' policy of 'one-sided disarmament'. The Liberal/SDP Alliance was equally critical of Labour's unilateralism.[10] Despite some important differences between the Liberals and the SDP over cruise missile deployment, the Alliance managed to present a coherent alternative defence policy to that of the government based on traditional themes. The Alliance opposed Trident but favoured a continuing nuclear capability for Britain. They confirmed their support for NATO but urged a reform of alliance strategy which would put less emphasis on nuclear weapons. In so doing they sought a middle way between what they saw as the Conservative's 'escalation of the arms race' and Labour's 'one-sided disarmament'.

The victory for the Conservatives in the 1983 election seemed to suggest that although bipartisanship has broken down there was still widespread support amongst the general public for traditional defence policies. In the post-mortem which occurred after the election it was generally acknowledged that defence was one of the issues which had been very damaging for Labour. One of the most strident condemnations of Labour's defence policy came from within the party from ex-Prime Minister James Callaghan. He had criticised the policies before the election and subsequently found himself under attack from some members of the party who believed that his comments had contributed to the divided image of the party during the Election. Callaghan's reply went to the heart of the matter:

> What the movement has failed to understand is that it reversed the traditional policy of the Labour party on which we fought 11 successive elections without any real attempt to convince the British people that what we were doing was right. I happen to believe it is wrong. But you make a fundamental mistake by believing that by going on marches and passing resolutions without any attempt to tell the British people what the consequences were, you could carry their vote. And you lost millions of votes.[11]

Callaghan's warnings, however, went largely unheeded. Within the party the view prevailed that it was not the message that was wrong but the way it was presented. After the election Michael Foot was replaced as leader of the party by Neil Kinnock, another unilateralist, and between 1983 and 1987 the party retained its commitment to a non-nuclear defence policy. In the 1984 National Executive Committee (NEC) document, *Defence and Security for Britain*, a strong effort was made to try to convince the public that the image that Labour was 'soft on defence' was wrong.[12] The document emphasised Labour's commitment to NATO and to improving Britain's conventional capabilities. In developing a new approach to defence the party also made strenuous efforts to try to avoid the impression of major divisions between Labour leaders which had been so damaging in 1983. Despite this attempt to give the party a more patriotic image, the opinion polls in the mid-1980s continued to indicate the unpopularity of Labour's approach to defence.[13]

While the Labour Party struggled to make its non-nuclear policies more acceptable to the public, the Liberal/SDP Alliance had its own problems with the defence issue. In 1985, the SDP produced a document entitled *Defence and Disarmament: Peace and Security* which emphasised the party's opposition to unilateralism.[14] The Labour Party was criticised for its promise to close down American nuclear bases in Britain which, it was argued, would bring about 'the most profound change in British defence policy in half a century'. According to the SDP this 'could lead to the withdrawal of the American commitment to the defence of Western Europe and the break-up of NATO'.[15] The SDP's staunch commitment to the British nuclear deterrent and multilateralism expressed in the document, however, caused difficulties with a section of the Liberal party. Liberal CND had always been a force in the party but from 1984 onwards they began to assert themselves more effectively, especially at the annual conferences. Although the Liberal leadership under David Steel attempted to limit the damage, resolutions opposing cruise missile deployment in Britain as well as any replacement for Polaris increasingly caused frictions within the Alliance during the mid-1980s.[16]

In an attempt to hammer out a common defence policy for the next election a Liberal/SDP joint commission on defence was set up in 1986. Despite large areas of agreement, the compromises reached on cruise missile deployment and the question of a replacement for Polaris did little more than paper over the cracks.[17] In the 1986 Liberal conference, the fundamental differences between the two parties were exposed when an amendment was passed supporting a non-nuclear defence policy.[18] In the aftermath of the conference both David Owen and David Steel emphasised their agreement on the need to maintain a minimum nuclear deterrent and attempted to re-establish a common defence policy for the Alliance – but the damage had been done. Like the Labour party in 1983 they went into the 1987 election with a widespread public perception that they were split on defence. Given the more general credibility problem facing the two parties (with two leaders) attempting to fight on a common platform, this division on defence proved to be a serious liability during the campaign. Accusations, especially from the Conservatives, that their defence policy was 'little more than a fudge' and that fundamental differences continued to exist did little to help their political fortunes in the election.[19]

Mrs Thatcher's third election victory in a row, in 1987, once again suggested that amongst the general public the post-war consensus on defence still remained reasonably firm, even if the inter-party consensus had seriously broken down. The Conservative Party alone remained unambiguously committed to the continuation of the Trident programme, a strong NATO, close relations with the United States and multilateralism. The unpopularity of Labour's policies and the division within the Liberal/SDP Alliance over defence during the campaign helped to deflect attention away from the problems which were increasingly facing the government's own defence policy. With the decision to discontinue the commitment to the 3 per cent real increase in defence spending, the difficulties of funding all the government's defence plans were becoming more and more apparent. These, though, were problems for the future and the obvious and immediate weaknesses of the alternative defence policies put forward by Labour and the Liberal/SDP Alliance must have appeared much more serious to the general public.

The poor showing of the Liberal/SDP Alliance in the election, and the subsequent split between David Owen and David Steel on the question of a merger, resulted in a further weakening in the political fortunes of the two centre parties. In contrast, despite its defeat, the Labour Party emerged from the campaign in a stronger position than in 1983. In the post-mortem which followed, the contribution which the party's defence policy had played in the defeat once again became a matter of controversy. The leadership announced a major review of all policies, including defence, but attempts by the centre-right to bring about a swing away from unilateralism were successfully opposed by the left wing at the 1987 conference.[20] Resolutions confirming the non-nuclear policy were passed and a number of speakers, like Ken Livingstone, warned the leadership not to give up the commitment to unilateralism.[21]

This controversy reappeared once again a few months later when Mr Kinnock attempted to modify party policy, firstly by accepting the idea of the American nuclear guarantee which he had previously rejected. Then in a television interview on 5 June he seemed to abandon the notion of unilateral nuclear disarmament (which he had traditionally supported).[22] This apparent shift in Labour Party defence policy resulted in the dramatic resignation of the party's defence spokesman, Denzil Davies, and a barrage of criticism from certain powerful sections of the Trade Union movement and the left wing. As a result of this hostility, Mr Kinnock was forced to retreat from his attempt to modify Labour's defence policy. In a subsequent interview with *The Independent* he indicated that party policy had not in fact changed at all.[23] At the end of the first round of the Labour party's policy review, therefore, despite the leader's attempt to make Labour's defence policy more acceptable to the general public, the party remained committed to unilateralism and a non-nuclear defence policy. Neil Kinnock continued to argue that nuclear weapons could not 'contribute to the effective defence of our country or the collective security of nations'.[24] The spectacle of the Labour Party leader doing a U-turn on the controversial question of unilateralism when faced with opposition from the left was not only damaging to Mr Kinnock personally, but it also exacerbated tensions and divisions within the party. Whatever the ultimate outcome of the review process, in the run-up to the next election it remains clear that the debate between the unilateralist and multilateralist wings of the party is likely to be protracted and bitter.

The political challenge to the traditional framework of British defence policy will probably continue for some time to come. Should the Labour Party maintain its non-nuclear stance and win the next election, in the early 1990s, the effect on British defence policy could be profound. The attempt to introduce the kind of policies which the party advocated in 1983 and 1987 would affect every part of the defence establishment in Britain. American nuclear bases could be closed down, defence expenditure might be reduced and attempts made to alter future procurement towards non-provocative weapons systems. This would require fundamental changes in the weapons, organisation and tactics of all three Services. Whether BAOR could retain its current role on the Central Front and whether Britain could retain its present command roles in NATO would be doubtful. How the NATO alliance and particularly the United States would react to these major changes is difficult to forecast. However, it does seem likely that the process of adjustment within the armed forces in Britain and within NATO would be painful.

Despite the U-turn made by Mr Kinnock in June 1988, on the question of unilateralism it is at least conceivable that the Labour Party's review process will arrive at some sort of compromise between the unilateralists and the multilateralists in line with the 'new realism' and the party's desperate need not to lose a fourth General Election in a row. Whatever form this compromise might take, it is unlikely that the Labour Party will be prepared to provide unambiguous support for Trident (although it might be maintained in the short term and included in the arms control process). It might be possible to fudge the questions associated with defence expenditure, US nuclear bases in Britain, and arms control. The continuing commitment to a non-nuclear defence policy, however, seems likely to remain. Such a policy, if maintained, would continue to represent a major challenge to the central pillar of post-war British defence policy and would prevent the creation of a new inter-party consensus on defence.

Another challenge to the Conservative government's defence posture may also come from the new Social and Liberal Democratic Party (SLD) which resulted from the merger of the Liberals and part of the SDP in early 1988. Without David Owen's commitment to the traditional elements of post-war policy, the new party is likely to be even more equivocal on nuclear matters than the old SDP/Liberal Alliance. The party seems likely to oppose Trident and will probably seek to negotiate it away in a future arms control agreement. On this issue at least a sort of limited inter-party consensus might emerge with the Labour party. The basic divide between the two main opposition parties and the Conservatives is likely to remain a key feature of the defence debate into the 1990s. In this respect, despite some differences, the fourth party, the SDP, may well be closer to the Conservatives than the other opposition parties. Certainly under Dr Owen's leadership the SDP is likely to maintain its basic support for the traditional tenets of British defence policy.

Predicting where the parties will stand on defence at the next election is clearly extremely hazardous. A great deal will depend on world events. Given the rapidly changing nature of East–West relations since Mr Gorbachev came to office, the international system may well be very different by the early 1990s. Even the most optimistic assessments of the arms control process suggest that radical cuts in strategic nuclear forces are unlikely to be achieved in full in the period before the next election, even if a START agreement is concluded. Despite the Conservative government's commitment to the American–Soviet START negotiations there has been a singular reluctance to include Britain's nuclear weapons in the process until substantial cuts in superpower arsenals have taken place. Even 50 per cent reductions would only bring the levels of strategic nuclear weapons down to the level of the mid-1970s. Whether the government would be prepared to participate in the arms control negotiations on strategic nuclear forces if the process gathered momentum remains to be seen. The commitment to Trident gives it some room for manoeuvre but there are clearly limitations on how much can be given up if a minimum nuclear deterrent is to be maintained.

How the government responds to the arms control challenge depends, in part, on public attitudes. Throughout the post-war period a favourable consensus towards nuclear deterrence has been maintained. Whether this consensus would continue in a period of sustained *détente* is difficult to assess. Improved East–West relations and a series of arms control agreements between the superpowers could bring growing pressure on the government in the early 1990s to include Britain's nuclear forces in the negotiations.

The INF agreement in December 1987 provides an interesting example of the challenge which arms control and better relations with the Soviet Union can bring. Although Mrs Thatcher's government supported the final agreement reached, there was clearly some concern in government circles over the impact on public attitudes towards European security.[25] In a speech in Bonn, in October 1987, Mrs Thatcher went out of her way to dampen down euphoria over the agreements and to warn against the view that the INF agreement was the first step towards a non-nuclear Europe. She was at pains to indicate that nuclear weapons remained an important ingredient of NATO's deterrent strategy. The government's task in this respect was not made any easier by the apparent willingness of President Reagan at the Reykjavik summit with Mikhail Gorbachev to work towards the elimination of all nuclear weapons by the end of the century. For Mrs Thatcher this was an unrealistic and dangerous aspiration which could only encourage false hopes and give sustenance to the anti-nuclear movement.[26]

The challenge posed by better East–West relations and the momentum of the arms control process was also demonstrated by the NATO summit meeting in March 1988. Although the meeting was held to try to demonstrate the cohesion of the Western alliance in the face of the Soviet 'peace offensive', it helped in fact to highlight important differences amongst NATO members over the modernisation of short-range nuclear weapons. Differences surfaced in particular between Chancellor Kohl of Germany and Mrs Thatcher. The German government wanted to pursue negotiations with the Soviet Union to achieve a 'third zero' agreement which would eliminate battlefield nuclear weapons from Europe as a follow-up to the 'double zero' agreement of December 1987. Mrs Thatcher, however, opposed this because of her concern that it would lead to a denuclearised Europe. Such a Europe, she believed, would be less stable because of the continuing superiority of Soviet conventional forces. Although the disagreement was papered over at the summit meeting, the potential for conflicts between NATO partners over how to respond to Soviet arms control initiatives was clearly evident.[27] As the process continues this challenge is likely to re-emerge.

This problem is part of a wider difficulty for Western leaders of how to react to the 'new thinking' coming from Moscow. The fresher, more conciliatory approach of Mr Gorbachev provides great opportunities for a period of sustained *détente* and cooperation between East and West after the hostility of the early 1980s. At the same time, Western leaders face a dilemma. Rapprochement may well lead to a growing unwillingness of Western publics to pay for effective defence. A hard line which emphasises the continuation of the Soviet threat, and which cautions that Gorbachev may not survive, may encourage the Soviet leader's more conservative opponents at home and thereby not only limit his welcome reforms in both domestic and foreign policies but could actually contribute to his downfall by denying him the success he needs to continue his reform programme.

This dilemma was clearly reflected in the behaviour of Mrs Thatcher towards the Soviet Union in late 1987 and early 1988. Initially, Mrs Thatcher let it be known that she felt Mr Gorbachev was a man she could do business with. Following Mr Gorbachev's stop-over at Brize Norton on his way to the Washington summit, in December 1987, British officials claimed a new special relationship was developing between the Soviet and British leaders. By March 1988, however, Mrs Thatcher's tone had changed and she was warning that the Soviet 'bear' was still a threat.[28] The Prime Minister appeared to be growing increasingly anxious that Gorbachev's peace initiative was having such an effect in the West that the public would be much less willing to sustain the necessary expenditure for defence. Maintaining effective defence and sustaining alliance cohesion is likely to remain a major challenge as long as Gorbachev's new approach to foreign policy continues.

There are other changes in the international system which also pose challenges for the future. Perhaps the most important of these stems from the relative decline in the power of the United States and the Soviet Union and increased pluralism in the international community. Although the superpowers remain by far the most powerful actors on the world stage, it has been clear for some time that their power has been declining. The American humiliation in Vietnam, the frustration of the Iran hostage crisis, the precipitate withdrawal from the Lebanon and declining economic strength (reflected in the huge budget deficit) all suggest that the United States is no longer as predominant in the international system as it once was.[29] The Soviet Union has also suffered serious setbacks in the Middle East, in its relations with China and especially in Afghanistan. Apart from these difficulties the major thrust behind Mr Gorbachev's new approach to foreign policy appears to be the urgent requirement for the restructuring (*perestroika*) of the Soviet economy. As a result the two superpowers appear to be moving towards some limited disengagement after the high profile foreign and defence policies which both have recently pursued. At the same time, other powerful actors have been emerging, including Japan, China and the EEC. More and more states have either acquired nuclear weapons or are close to producing such a capability, including India, Pakistan, South Africa, Israel, Brazil and Argentina. Bipolarity may not be over but there is a clear trend towards greater pluralism in the international system.

This growing diffusion of power in the world community clearly poses particular challenges for British statesmen. Bipolarity has provided a fairly simple map for the conduct of Britain's foreign and defence policy since 1945. As the world moves towards greater pluralism and the United States begins to devolve power (as seems likely), the tasks of British policy-makers are likely to become more demanding. Can the Atlantic Alliance continue in its present form? Is there any viable alternative? Should Britain continue to emphasise the 'special relationship' with the United States? Should more be done to contribute to European defence cooperation? Should Britain continue to encourage the process of rapprochement with the Soviet Union? Should greater emphasis be given to national as opposed to alliance defence capabilities? Such questions are likely to pose great challenges to those responsible for deciding on the right balance to strike in British defence policy in the years to come. How they should approach this task is the subject of the remaining two chapters of this book.

9 Incrementalism Versus a Radical Review

The incrementalist approach to British defence policy-making since the Second World War has reflected the determination of successive governments to maintain a balanced force structure. Although Britain has long since ceased to regard herself as a Great Power on a par with the United States and the Soviet Union, the provision of a wide range of military capabilities has been considered necessary to sustain the nation's self-image as a major medium-range power. Writing in 1982 David Greenwood argued:

> the main characteristic of the present day national order of battle is 'balance' in both inter-service and intra-service senses. A small strategic nuclear force is complemented by comprehensive conventional capabilities in all three branches of the armed forces, which continue to receive near-equal shares of annual appropriations. The Royal Navy no longer has attack carriers or a fully-fledged amphibious capability, but in all other respects the United Kingdom retains a balanced fleet. The army has recently undergone 'restructuring' but remains a true all-arms force. Short of combat planes and fast jet pilots though it may be, the Royal Air Force has, and plans to continue to have, the wherewithal to perform the full spectrum of tactical air missions.[1]

This assessment of Britain's defence capabilities remains largely valid in the late 1980s. Britain is still making an important and 'high-quality contribution to NATO's order of battle at sea, on land and in the air; in all three elements of the triad (conventional forces, theatre nuclear weapons, and strategic nuclear capabilities); and to the ready forces in each of the alliance's major command areas (Atlantic, Channel and Europe)'.[2] Britain also retains residual intervention capabilities for operations outside Europe as well as garrisons in places like Belize, Hong Kong, Gibraltar and the Falkland Islands. An attempt at unbalancing Britain's defence effort in the 1981 review was thwarted by the Falklands War and ever since the government has attempted to maintain the traditional commitment to comprehensive and balanced military capabilities.

But how long can this last? As the last three chapters have tried to show Britain faces a wide range of economic, political and technological challenges in the future which are likely to force the government to look once again at the balanced force concept. For supporters of incrementalism and the flexibility which balanced forces provide there is little need for significant changes in direction. If cuts have to be made adjustments should be made across the board as painlessly as possible. There are other commentators, however, who argue that British defence policy is approaching an important crossroads in the late eighties and early nineties.[3] They argue that, given the pressures, the incrementalist approach will inevitably result in an ever weakening defence effort. What is required is a radical review which will focus on priorities and allow Britain to specialize in those areas which are most vital to British security. Who is right? Before we can decide we need to look at the arguments in more detail.

In 1985 the House of Commons Defence Committee undertook a review of Britain's current and future defence commitments and resources in the light of the government's decision to end the 3 per cent real increase in defence spending.[4] The Committee suspected before it undertook the review that a policy based upon level funding (which in practice would lead to a decline in defence spending in real terms in the late 1980s and early 1990s) would be likely to create an imbalance between commitments and capabilities. Having taken evidence from the Ministry of Defence and the Services, the Defence Committee report concluded that these fears had not been allayed. They were left with 'the strongest suspicion' that future defence budgets would fall 'substantially short of the resources required to maintain capability and meet commitments'.[5] The arguments presented by the Permanent Under Secretary, Sir Clive Whitmore, that improved efficiency, competition, collaboration and flexible planning would substantially resolve the difficulties were found unconvincing by the Committee. According to the report the kinds of savings produced by such a 'value for money' campaign would simply not be enough. It was felt that the 'funding gap' produced by a defence budget which is falling in real terms would be too wide to deal with using such remedies.[6]

This view that the government is going to face some hard decisions in the defence field in the next few years is widely held by defence analysts, opposition politicians, former Cabinet Ministers and the press. On 12 January 1986, a leader article in *The Sunday Times* concluded that 'the arithmetic of future spending shows that unless there is a radical rethink, matters can only get worse'.[7] Much the same argument was presented by John Nott in two articles he wrote for *The Times* in October 1987.[8] According to the former Conservative Defence Minister it was 'hard to see how it will be possible to keep up the main thrust of the current investment programme without increasingly awkward political decisions'.[9] There can be no doubt that this assessment is correct.

It has already been argued that, faced with these awkward political decisions, the government has three main options. One is significantly to increase the resources allocated to defence. The second is to undertake a radical review, establish priorities and cut one of the major commitments. The third is to pursue the traditional policy of incrementalism, or 'muddling through' seeking better value for money in every way it can but accepting that a range of smaller cuts together with slippage in procurement dates will be necessary.

The first of these options is possible but unlikely. It is true that the Labour and Conservative governments of the late 1970s and early 1980s did accept an annual increase of 3 per cent in real terms (from 1977 to 1986) after defence had been somewhat neglected in the mid-1970s. It is conceivable that the government, faced with damaging cuts in defence capabilities in the late 1980s and 1990s would once again respond in this way. Faced with having to make 'some difficult choices' in the autumn of 1988 the government did, after all, decide to spend a little more on defence in the period up to 1991/2. With the economy doing well and the traditional commitment of the Conservative government to strong defence this approach cannot be ruled out. It must be said, however, that the chances of *substantial* increases in defence spending towards the end of this decade and the beginning of the next are not very good. At a time when the government will be facing another General Election it probably won't want to transfer resources away from Education, the Health Service or Social Services. Its determination to continue with the campaign to keep public expenditure under control and maintain economic growth would also make major increases in defence spending unlikely. With the improvement in East–West relations in the late 1980s and Gorbachev's 'peace offensive', the government would also find it difficult to sell major increases in defence spending to the public.

That would seem to leave the two options of incrementalism or a radical review. In order to assess which of these approaches is preferable an analysis of the strengths and weaknesses of both approaches to decision-making is necessary.

The case for continuing the incrementalist approach is often based on a critique of the effects which a radical review, based on the cost-effective approach, would be likely to have on British defence policy.

Two criticisms, in particular, are often heard. Firstly, there is the argument that radical reviews like the one undertaken in 1981 have a harmful effect on British security.[10] According to this point of view, it might be rational from an economic perspective to establish clear priorities and focus sharply on certain areas of defence effort. But the effect of such a policy can be to weaken important areas of defence significantly and close off various options for the future. In 1974, it is argued, the government decided that there were four *vital* areas of defence policy. In 1981, however, it was decided to impose major cuts in one of these areas (the surface fleet role in the defence of the Eastern Atlantic). According to the critics of the 1981 defence review that was short-sighted and was shown to be short-sighted by the Falklands War of 1982 when the very ships which John Nott had decided to cut made an invaluable contribution to British victory.

The main thrust of this argument, putting aside the merits or otherwise of the specific decisions made in 1981, is that if the four areas identified in 1974 were indeed vital to British security every effort should have been made to sustain these commitments. The fact that some areas of defence are regarded as relatively more important than others does not mean that the least vital area is not of continuing importance to the security of the state. This problem was in fact recognised in the White Paper entitled *The Way Forward*, produced by John Nott in 1981. According to this the defence of the Eastern Atlantic was still of major importance. In the words of the White Paper: 'that such a contribution must continue, and on a major scale, is not in question'.[11] The government's decision, however, was to try to provide this contribution in a different way and more cheaply. For many senior naval officials the new emphasis on a different mix of forces was inadequate to do the job effectively. The government had accepted that the role continued to be crucial but it was not prepared to provide the resources or the capabilities to maintain effective defence forces.[12]

The danger of weakening important areas of defence capability, the critics argue, is inherent in the 'costed-options' approach. The problem is that it brings the economist's logic to bear in an area where politics and security require a different approach. Far better to introduce cuts here and there and accept some cancellations of

equipment and deferrals which may weaken the whole defence programme marginally, rather than undermining a major defence capability upon which the security of the state depends.

Another criticism of the radical review approach is that it is not in
tune with political realities.[13] According to this argument over the
post-war period as a whole, whenever radical changes in defence have
been proposed they have usually been whittled away as a result of the
political and bureaucratic processes which are characteristic of the
British policy style. This was true of the Sandys White Paper in 1957,
the 1974–5 defence review and the 1981 defence review. As John Nott
conceded in his article in *The Times*, on 5 October 1987, 'the obstacles
to a rational process are manifold' in the British system. He identified
three obstacles in particular. Firstly, the problem created by a system
of annual Treasury control which 'is so organized as to prevent any
serious long-term financial planning'. Secondly, the difficulties
created by the central role of the Services in the planning process.
According to Nott 'because of the disciplined and hierarchical nature
of Service life there is bound to be something of a negative influence
on any radical forward thinking'. The Services are not normally
'conceptual thinkers'. And thirdly, he argues that incrementalism is
encouraged by the roles of both civil servants and Ministers. John
Nott's experience of government suggested that inertia was built in to
the system of decision-making. Civil servants tend to see their main
function as keeping their Ministers out of trouble and the general
ambition of most Ministers is, understandably, to retain the support
of their party, including its Service lobbies. As a result it is not
surprising that there is a huge weighting in favour of traditional
approaches against any shifts in long-term direction. These obstacles
are obviously deeply ingrained in the British system of government.

Although John Nott himself attempted to introduce major changes
in the substance of policy in 1981, there is no doubt that his review
created a great deal of anxiety in the Navy and amongst Tory
backbenchers which the government found embarrassing. Those
defence analysts who argued that a major review would have to be
undertaken before the 1987 election failed to recognise that it was
highly unlikely that a Conservative government would wish to face
another bruising and politically damaging debate about defence in the
run up to an election.[14] Avoiding such internal divisions in the party
(which a major defence review would almost certainly have created)
was particularly important in 1987 because of the government's belief
that their Labour and Alliance opponents were weak and divided on

the question of defence. Even after the election there has been a very strong incentive to muddle through incrementally, adjusting policies at the margin rather than radically weaken one area of the nation's defence capabilities. In this sense incrementalism can be said to suit domestic political realities rather more painlessly than the radical review approach.

It is perhaps not surprising that incrementalism has had a bad name. After all 'muddling through' is not a very attractive description of policy-making. Incrementalism should not, however, be dismissed out of hand. Certain characteristics of this approach would appear to have had a beneficial effect on the conduct of British defence policy since 1945. The post-war period has been very much a time of transition for Britain as a constant process of adaptation to changing domestic and international circumstances has taken place. Future historians no doubt will view the relatively peaceful and smooth withdrawal from empire much more favourably than contemporary observers.[15] In such periods of major transition a policy process which helps to keep a range of options open and which encourages 'strategic pluralism' can well produce better results than one which focuses more narrowly on areas of key priority.[16] As we have seen, British policy-makers have attempted to strike the right balance between the continental commitment *and* a maritime strategy, European *and* overseas security, the 'special relationship' *and* European coopera- tion, nuclear *and* conventional forces, as well as alliance *and* national interests. With some justification successive governments recognised that the range of British political, economic and security interests would be better served from such a balanced approach than one which closed off a range of options and capabilities through forcing policy-makers to make stark choice on the basis of 'costed options'.

A case can certainly be made for 'strategic pluralism' as the basis of British defence policy in the late 1980s and 1990s. In the light of the political, economic and technological challenges facing British defence planners and all the uncertainties about the future, there is a strong argument in favour of retaining as wide a range of choice as possible. At a time when the structure of world power appears to be undergoing significant changes it would seem to make sense not to erode or eliminate important areas of defence capability which might prove to be of major importance in the future. In the late 1980s, the British government has been faced with a range of anxieties about such things as the future of the American commitment to Western Europe, the sincerity of Gorbachev's peace offensive as well as his

ability to retain power, the direction of future arms control negotiations, and the significance of greater Franco-German defence cooperation. Faced with such worries and uncertainties the case for
continuing to 'muddle through' may well appear to some decision-
makers less painful than facing hard decisions about priorities. There
is also the fact that clear choices, if they are the wrong choices, can be
disastrous.

Despite the value of incrementalism there are also weaknesses in
this approach to decision-making as well. One of the main criticisms
of British defence policy since 1945 is that it has suffered from a lack
of priorities. The general framework of defence laid down in the 1940s
has remained in place with the exception of the decision to withdraw
from east of Suez in 1967–8. Little innovative thinking has occurred
as the government has attempted to strike a balance and make
adjustments in the major areas of policy. The result has been inertia
and constant strain as commitments and capabilities have frequently
been out of line. Rather than long-term planning governments have
been forced to make hasty decisions in reaction to economic crises.
Although a number of attempts may have been made to identify clear
priorities in the defence field (for example in the 1974–5 and 1981
reviews), in practice these have been whittled away or simply over-
taken by events.

Faced with the range of challenges described in Chapters 6, 7 and 8,
a continuation of this approach to defence decision-making in the
future is likely to have serious consequences. According to John Nott,
'we simply have to change direction' because:

> Ultimately, a policy of seeking to be all things to all men and to be
> ready to meet every possible contingency can lead only to a
> degredation of every capability – not enough men, training, fuel,
> missiles or even ammunition. Everywhere in the Ministry of
> Defence, companies, individuals and institutions know that they
> must specialize to succeed ... Without the language of priorities,
> the application of scarce means to given ends, we face a bleak
> future for the security of the United Kingdom.[17]

This criticism that incrementalism leads to 'cheese-paring' all round
which ultimately weakens the whole of the defence programme is one
which is shared by many observers. In its campaign for a radical
review of defence *The Times*, on 12 January 1986, quoted an unidenti-
fied Senior Officer as saying that 'salami slicing' would reduce the

operational effectiveness of the armed forces 'below their already low levels'. The only answer, he said, was 'to take hard decisions about priorities'.[18]

This has also been the argument developed by one of the strongest proponents in Britain of the need for a radical review of defence, David Greenwood. Greenwood, a leading defence economist, has constantly campaigned for a more rational approach to defence decision-making. In his view the government's short-sighted approach to defence will not solve the longer-term problems. Like the other critics, he argues that it will weaken the whole of the defence effort. More than this, however, in his opinion, it will also result in a loss of democratic control. He has put the case for change this way:

> The 'value for money' crusade will not be enough . A fully-debated exercise in setting or re-setting defence priorities is preferable to the surreptitious massaging of plans. There are two reasons for this. First, proceeding by ad hoc 'adjustments to the programme' means in effect muddling through without addressing the issue of priorities. The result tends to be undernourishment across the board ... The second reason for favouring a thorough-going reappraisal of the defence effort is that ... when 'cash management' rules it is the Accountants who have the last word and important shifts in priorities can occur without appropriate political discussion.[19]

Such shifts in priorities represent a defence review of sorts but it is a 'defence review by stealth'.[20]

David Greenwood makes a powerful and convincing case for a more rigorous and systematic attempt to identify 'essential defence needs' within the context of the resources available. He has long been an advocate of the 'costed-options approach' pioneered by Hitch and McKean in the United States.[21] The costs of a variety of defence packages or options are worked out and Ministers choose which should be given priority. In contrast to the 'equal misery approach' which weakens defence across the board, the costed-options approach is designed to sustain (and improve) those areas of defence capability which are most vital to the security of the state while cutting back in those areas which are regarded of less importance.

Attractive as incrementalism may seem, the fact remains that insufficient resources are available to meet the country's defence commitments as defined by the government. In a period of little growth in defence spending incrementalism poses very real problems.

A wide range of military capabilities are likely to suffer. The effects of such a policy were graphically outlined to the House of Commons Defence Committee by the Chief of Defence Staff in 1986:

> There will definitely be slippage and in-service dates will move back and therefore capability gaps measured against the threat will obviously not be closed quite as quickly as they were before. Because the in-service dates get deferred, the old equipment, which new equipment would have been replacing, will have to run on a little longer and, of course, the capability gap between the opposition and yourself will therefore increase; their technology is improving and you are having to hold yours down. There may also be manpower implications, although not necessarily in all the Services. In the Navy, for instance, if you have to slip the in-service date of a modern frigate with a relatively much smaller ship's complement and you have to run on the older frigate, then of course that will produce manpower problems. Then it may be necessary to cut down on sophistication, and indeed this may not always be a bad thing – sometimes I feel our sophistication goes too far – but you have a very sophisticated threat and you might have to cut down on your sophisticated counters to it ... Also things like ammunition, which you were going to make good and bring up to full force goal requirement, you might not be able to do quite as quickly as you have done before. These are the areas where I think reductions may be considered and you have to measure that against (a) what it does to your deterrent posture and, (b) what it does to your war fighting capability.[22]

This assessment of the impact of incrementalism by the CDS gets to the nub of the problem. Such 'slippage' can allow Britain to continue with the strategic nuclear deterrent, to provide defence for the Home Base, the Central Front and the Eastern Atlantic and to retain some intervention forces. The difficulty is that Britain's total defence effort is likely to be weakened. Where the reductions take place will obviously reflect priorities but the overall effect will inevitably be a decline in Britain's armed forces more or less across the board. That might be acceptable in the short run – indeed the reductions identified by the CDS, worrying as they are, do not seem to be too drastic. In the long run, however, the effect would be very serious. A continuous process of ad hoc cuts would lead eventually to a point where Britain's armed forces would cease to be effective. Rather than reach such a point it is much more sensible to weaken or eliminate the least important component of defence policy in the wider interests of the total defence effort.

The dilemma which defence planners face is that identifying where this point is exactly is extremely difficult. Judgements about the irreducible minimum of defence effort will inevitably vary, even within the military establishment itself. All may agree that reductions will gradually weaken defence capabilities, but deciding the point at which further cuts will fundamentally undermine the integrity of the whole defence effort is likely to result in genuine differences of opinion. Given this difficulty the government has to try to balance the advantages associated with keeping its options open with the dangers of undermining a wide range of defence capabilities. It also has to balance the advantages of establishing priorities and long-term planning against the dangers of eroding useful military capabilities which could, in certain unforeseen circumstances, prove invaluable in the defence of national interests.

There is clearly no easy way out of this dilemma. The incrementalist and best value approaches both have strengths and weaknesses. By far the greatest danger, however, is that 'the muddling through' process will weaken defence capabilities across the board. It may well be that incrementalism with its advantages of keeping as many options open as possible has suited Britain's defence needs in the past. The contemporary pressures on defence suggest that unless substantial extra resources are made available the maintenance of effective balanced forces will become increasingly difficult. John Nott would seem to be correct in his argument that, 'Until recently it was possible to go for balanced forces ready to meet whatever crisis faced the country, but that luxury is no longer available'.[23] The fact that the post-war period has been punctuated by a series of defence reviews at irregular intervals conducted in different ways, using different criteria and usually in haste in order to resolve an immediate budgetary crisis, reflects the short-term focus of the incrementalist approach to defence policy. Such incrementalism is probably to some extent inevitable, not only because it reflects the British tradition and domestic political realities but it also serves important national needs of preserving a spectrum of defence capabilities to deal with uncertain events in the future. Nevertheless, the need for longer-term vision in defence planning and a more systematic evaluation of alternative means and ends is now so urgent that some form of review process is required. But what form should this take?

Periodic radical reviews which have been used in the past have often been highly controversial in political terms and have rarely had a long-term effect on defence planning. It would seem sensibile, therefore, to build into the present system a regular defence review, every five years as they do in the United States and France. The American Defense Guidance Plan and the French *loi de programmation* are designed to provide long-term strategic direction for defence planners.[24] Such direction is badly needed in British defence policy. The former Defence Secretary, John Nott, provided a damaging indictment of the weaknesses of the existing system to undertake long-term planning when he described the Ministry of Defence as being 'like a huge super-tanker, well captained, well-engineered, well-crewed, its systems continuously up dated – but with no one ever asking where the hell it is going'.[24] What a regular defence review process could do would be to provide a sense of direction and a systematic periodic assessment, in the light of changing domestic and international circumstances, as to whether the best defence at the lowest cost was being achieved.

Any review of defence is bound to be an occasion for controversy but in the past political differences have been heightened by the manner in which reviews have been undertaken. There has been a tendency to try to avoid them for as long as possible and when they have taken place they have usually been forced on the government in a situation of crisis. It would seem to be far better to accept the value of a regular defence review process which would help to overcome the politically charged atmosphere of past reviews and allow cooler, more considered assessments to be made of the effectiveness of the defence effort.

The author first put forward the idea of a regular defence review in a memorandum to the House of Commons Defence Committee in March 1988. An almost identical proposal was subsequently put forward in June 1988, in a paper written by Leon Brittan. The former Secretary of State for Trade and Industry argued that

> the changes currently taking place and those in prospect must surely lead us to take a fresh look at the pattern of British commitments and our ability to meet them effectively. The logical way of doing this is to have a defence review ... It cannot be right to muddle through if there are legitimate grounds for a review ... A defence review should not therefore be regarded as something to be dreaded, and only to be contemplated if tomorrow's bills cannot

be met. It should, instead, be perceived as a rational step to take from time to time, when it looks as if political, strategic and financial circumstances may have changed ... As threats, commitments and force structures change, a review asks two fundamental questions: Where do we go? How do we want to get there? It considers if cuts are necessary, or more resources required, if priorities are correct, if spending should be reapportioned, if resources are being used effectively, if value for money is being obtained. Defence reviews involve fundamental issues of foreign policy and judgements on military strategy. The key element in winning support for a review lies in taking away the political significance of calling one. By having a regular procedure of fundamental, external reviews every 4–5 years, governments would be able to take a more strategic view than would otherwise be possible.[25]

Although Brittan was no longer in the Cabinet it was significant that such a high-placed Conservative politician should support the call for a regular process of defence reviews, especially as the government itself continued to argue that a review of any sort was unnecessary.

It is clear that the introduction of a regular defence review process and the adoption of a five-year plan would not be without difficulties. It would require some adjustment within the Ministry of Defence, especially in terms of the longer-term planning required by weapons' procurement.[26] These structural changes, however, would become manageable if the government became more committed than it is at present to the idea of providing more effective longer-term strategic direction for defence policy as a whole. At present the Central Policy Staff, which on paper is supposed to provide an important input into future planning, is badly understaffed and in practice spends its time on more immediate problem-solving.[27] The expansion of the Central Policy Staff and the introduction of a bureaucratic framework for regular strategic assessments of where the defence effort 'is going' would make a very useful contribution to the future coherence and effectiveness of British defence policy.

The argument that five-year plans would create a 'strait-jacket' effect and significantly weaken the flexibility of the defence effort to deal with unforeseen events is not very convincing. The idea would be to use the five-year plans (as both the Americans and the French do) as a loose but coherent framework to provide guidance and direction for defence planning as a whole. Incremental changes and a degree of flexibility can still be accommodated within the broad strategic direction provided by the periodic defence reviews and the future plans which result from a review process.

Simply changing the structure of defence decision-making and attempting to improve longer-term planning does not, of course, solve the problems of future British defence policy. Neither does it necessarily resolve the tensions between the incrementalist and best value approaches to defence decision-making. Whether the government decides to keep its options open and strike a balance between the different areas of defence policy, or whether it decides to be more systematic about the priorities and specialise in certain areas, is ultimately a matter of political judgement. Just as the question of how much to spend on defence cannot be answered in any definitive way, so the question of how to allocate resources between different roles is likely to be the subject of endless genuine differences between the Services themselves, defence planners and outside commentators. As we have seen in the first five chapters of this book, the continuous attempts by successive governments to strike what they regard as the 'right' balance in a number of key areas of defence policy have been, and remain, the subject of great controversy and disagreement. Defining what exactly is the 'right' balance is not a matter of objective strategic analysis, it is the result of value judgements. Whether the continental commitment should be given more priority than the maritime strategy, European defence more priority than overseas defence, the 'special relationship' priority over European cooperation, nuclear weapons modernisation more priority than conventional weapons modernisation, and national defence requirements more priority than alliance defence requirements, are questions which are not susceptible to right or wrong answers. A multiplicity of answers, all of them 'right', can be offered. In such circumstances, as W. R. Schilling has argued, the opportunity for reasoned and intelligent conflict with regard to the factual premises involved are legion. The questions of value involved are, in the final analysis, matters of personal preference.[28] As such it is inevitable that there will be differences between 'good, intelligent and dedicated men' about the foreign policy goals of the state, about the relative effectiveness of the various means available to achieve those goals, as well as about how much to spend on defence and all the non-defence opportunity costs involved in such expenditure. Ultimately, choices between values have to be made and such choices can only occur in one place: the political arena. It is there that the relative importance of values can be decided

by the relative power brought to bear on their behalf. It is only in the
political arena that 'the distribution of power can decide matters that
the distribution of fact and insight cannot'.[29]

The fact that questions of value inherent in decisions about defence
are, in the final analysis, matters of personal preference and that, as a
result, defence policy must be a matter of political choice does not
mean that rational analysis plays no part in the process. What the
introduction of a regular review process can help to bring about is a
recognition that priorities have to be established in the context of
clearer strategic direction. Rather than 'muddling through' and
weakening defence across the board some form of specialisation is
going to have to take place. The concept of 'balanced forces' in the
context of the problems raised by a more or less static defence
budget, defence inflation and technological change can only lead to
a gradual decline in Britain's defence capabilities. The first part
of the regular review process proposed here should be undertaken as
soon as possible. After years of arguing that a review was unnecessary,
the Defence Secretary, George Younger, accepted in February 1988,
under pressure from the Chiefs of Staff, that crucial defence
programmes would have to be cut unless more money was forthcoming
from the government.[30] Once again, in June 1988, it was reported that
the Defence Secretary had told the Chief Secretary to the Treasury,
John Major, that a substantial increase in the defence budget would
be necessary to avoid 'a fundamental review of Britain's defence
commitments'.[31] More money was made available for defence in
November 1988 but not, it would seem, on the scale likely to bridge the
'funding gap'.[32]

Even if further resources are allocated to defence in the next few
years, the need to reassess the total defence effort in the light of
contemporary challenges and the need to provide long-term direction
is certainly necessary. If such funds are not forthcoming (which is
likely at least on the scale necessary to fill the 'funding gap'), then the
requirement for such a review which focuses on priorities and accepts
the need for more specialisation in the defence field becomes even
more urgent. What form such a review might take is the subject of the
last chapter.

10 Striking the Right Balance for the Future

British defence policy is clearly at an important crossroads. The argument presented in the last chapter suggests that the government cannot go on for much longer, as it has been doing in recent years, 'muddling through' on a day-to-day basis without serious damage to the nation's defence effort. A significant increase in defence expenditure would help to ease the immediate problems created by a 'funding gap', but as we have argued earlier, this appears unlikely to happen. If the government is not to weaken defence across the board by incremental 'salami slicing', a more radical assessment of priorities is clearly necessary. But what form should this take?

One of the important contemporary events which affects British defence policy is the growing ambiguity of the Soviet threat. Assessments of threat are always notoriously difficult given the problem of weighing intentions and capabilities. The problem has become even more difficult as a result of Gorbachev's 'new thinking' and the impact which this is having on Soviet domestic politics and foreign policy.[1] Ever since the Soviet General Secretary surprised Western leaders in 1985 by accepting the American proposal for eliminating intermediate range missiles in Europe, Western capitals have been in a state of confusion. No one expected the Soviet Union to accept the idea of eliminating their short-range and medium-range missiles in exchange for the elimination of the West's cruise and Pershing missiles deployed in Europe. As a result very little thinking had been done about the implications of the agreements for NATO strategy, or for future arms control policy.[2] Indeed, very little consideration had been given, in general, to the impact of the changing nature of East–West relations on the future of British defence policy. Therefore, one of the first tasks of the review process (which is just as urgent for NATO as a whole as it is for Britain) must be to assess defence requirements in the light of the changing structure of international politics. In particular the question which must be asked is whether there is still a Soviet threat which necessitates the *kind* of defence posture Britain has sustained for most of the post-war period?

There is certainly a strong case for welcoming the 'new thinking' in

118

Moscow – even though this complicates the task of governments in justifying their defence policies. Do we really prefer a hostile aggressive Soviet Union to one which is behaving more responsibly in the international system, and one that appears to be supporting ideas that we ourselves have promoted for many years? The chance of being able to reduce significantly not only the numbers of intermediate range missiles but also strategic, conventional and chemical weapons is surely something which should be welcomed by the West. Security depends on political relations as well as military balances. Greater cooperation and an easing of tensions can do a great deal to improve Western security.[3]

At the same time, Britain and her allies have to be rather cautious about the new ideas emanating from Moscow. We do not, of course, know whether Mr Gorbachev will survive. His domestic reforms after all are likely to give rise to growing expectations which will be exceedingly difficult to control. As we have seen already with the sacking of Mr Yeltsin, the Moscow Party boss, there are other political (and maybe military) figures waiting in the wings with less enlightened views than Mr Gorbachev appears to have, who remain suspicious of *glasnost* and *perestroika* at home and common security abroad. Even if Mr Gorbachev does survive, the Soviet Union remains a great power with immense military strength. Soviet leaders in the past have shown a predilection for using their military capabilities to resolve difficulties around their borders. No British or Western military planner can therefore conclude with certainty that the Soviet Union poses no threat to our interests. While the process of dialogue must continue and while we should be positive about the importance of seeking mutually beneficial arms control agreements, the British government and its allies must sustain an effective deterrent and defence posture.

How Britain can manage to sustain an effective deterrent and defence posture in the context of the limited resources devoted to defence should be the key question of any review process. As in 1981, each of the four and a half priorities should be once again scrutinised with great care. The key question is whether the same priorities should be established again.

THE STRATEGIC NUCLEAR DETERRENT

The arguments for and against nuclear deterrence in general and the

British nuclear deterrent in particular have been rehearsed over and over again in numerous studies in recent years.[4] For Britain, the fundamental question is whether, in an unpredictable world, nuclear weapons enhance British security or not. This is clearly a very difficult question to answer. While moral questions may be important ultimately it is a matter of judgement about the balance of risks.

The dilemma at the heart of deterrence theory is very troubling for any sensible person. Peace is preserved through the threat of a nuclear response which, if carried out, could conceivably lead to the end of civilisation as we know it. From that dilemma comes both the strength and the weakness of deterrence. The strength is that faced with such cataclysmic consequences most rational decision-makers would refrain from aggression. Nuclear deterrence seems to work. The fact that a traditionally war-torn Europe has remained at peace since 1945 and that one of the major contrasts between Europe before 1945 and after is the existence of nuclear weapons seems to give some credence to this idea. The weakness of nuclear deterrence theory, on the other hand, is that it depends on rational calculations which may not always be present in crisis situations. It also tends to neglect the consequences if deterrence breaks down. This could happen not only as a result of irrational action but also because of accidents or miscalculation.

Clearly no one can be complacent about the risks of living in a nuclear world. Nuclear weapons have been invented, however, and cannot be disinvented. Even if it was possible to achieve Mr Gorbachev's and President Reagan's declared goal of eliminating all nuclear weapons, the knowledge about how to produce them would remain.[5] And the fact that such knowledge could be used by ruthless leaders in the future means that it is pointless to try to turn the clock back. We are going to have to continue, whether we like it or not, to live in a nuclear world, irrespective of what President Reagan and General Secretary Gorbachev have said about wanting to rid the world of nuclear weapons.

While the Soviet Union continues to possess nuclear weapons Britain has the choice (as long as it remains in the Atlantic Alliance) of relying on others for nuclear protection or remaining a nuclear power herself.

The problem with relying on other states is that vital interests might in the last resort prevent allies from fulfilling their nuclear guarantees. In most circumstances the deterrent effect of extended deterrence might work. Past history, however, suggests that states are quite capable of letting their allies down in the pursuit of their own interests. Suez remains a reminder that even in a relationship as close as that between Great Britain and the United States, American Presidents can act in a way which significantly undermines the perceived interests of their ally across the Atlantic. Can Britain be sure, especially in an age of strategic parity and growing superpower disengagement, that the United States would always shelter Britain under its nuclear umbrella? Despite the often repeated rhetoric that an attack on Chicago would be regarded as an attack on Munich or Birmingham, the European allies can never be certain that when the time comes the United States would always react in this way. To be sure, at present neither can the Soviet Union be one hundred per cent positive that the US would *not* launch its missiles on behalf of Western Europe. Even though this would not be a wholly rational action, the Soviet leadership cannot be certain. But will the Western alliance always remain as it is today? Will the United States always perceive its interests to be so closely entwined with those of Western Europe? Already there are signs that the United States is beginning to refocus its foreign policy more towards Central America and the Pacific. The publication of the Pentagon's *Discriminate Deterrence* report in early 1988 indicated a tendency to downgrade the priority given to Western Europe within the American defence establishment.[6] Because logic suggests the US would not commit suicide on behalf of its Western European allies and events indicate a reorientation of American foreign and defence policy, if not disengagement, the case for a nuclear insurance policy is a strong one. In a world which remains dangerous and unpredictable there is a good case to be made that nuclear weapons are reassuring, and they just might be decisive at some point in the future in preserving the British identity, values and independence. As the 1975 House of Commons Expenditure Committee report argued:

> In the last resort, if the Alliance was to collapse, the possession of an independent strategic weapon provides the UK with a means of preserving national security by deterring large-scale conventional or nuclear attack or countering blackmail.[7]

This may be a Gaullist view but it's no less convincing for that.

The arguments which opponents put forward against this view are far from convincing. It is often said that the circumstances in which the Alliance might collapse and Britain would have to stand on its own are so inconceivable as to be not worth planning for. It is true that it is difficult to imagine the precise circumstances in which Britain might have to stand up against the Soviet Union. This does not mean, however, that such circumstances can be dismissed. Remote contingencies (especially if they are crucial to the ultimate survival of the state) have to be catered for, provided the economic resources available allow it. As we have seen many times in the past unlikely contingencies do happen from time to time. If the consequences of not paying the premium for such an insurance might be high (as it would be if Britain ever confronted another nuclear power alone – whether it be the Soviet Union or any other nuclear state), then there would seem to be a strong case for prudence in defence decision-making.

Another criticism is that even if such circumstances are conceivable, a British nuclear deterrent facing the overwhelming power of the Soviet Union is bound to be incredible. After all, the use of the British deterrent would almost certainly bring about retaliation which would destroy the British way of life. Such knowledge of the relative inequality of the destructive power of Britain and Soviet nuclear capabilities, and the unequal consequences of their use, would prevent any sane British leader from initiating an attack and the Soviet leaders would know this. Britain could destroy a number of Soviet cities but the Soviet Union could wipe Britain off the face of the earth. Both sides are aware of this and Britain would be self-deterred, so the argument goes. Such arguments are not as convincing as they first appear. What matters is what the *Soviet* perception would be in a situation without precedent and of unique peril. Could they guarantee with absolute certainty that British leaders would not launch an attack? Could the Soviet leadership (or any other) ever discount the possibility that, faced with a threat to a way of life built up so proudly over many generations and defended so resolutely and at such cost in the past, Britain might fire its missiles? Small as the British nuclear deterrent is in comparison with that of the Soviet Union, it is nevertheless capable of inflicting truly horrendous devastation. The Soviet Union could not afford to get it wrong. Even a 5 per cent chance of British retaliation, therefore, is likely to be extremely effective. As Kenneth Waltz has argued, the existence of nuclear weapons tends to produce great caution in the behaviour of the nuclear powers.[8]

There are those who suggest that these arguments in favour of a British nuclear deterrent apply to other states as well. Britain, it is argued, is thus guilty of hypocrisy in trying to prevent nuclear proliferation. Indeed, the British example is likely to encourage other states to acquire nuclear weapons. In a sense, the first part of this criticism is correct. British policy is somewhat hypocritical in this respect. Britain, however, has an obvious interest in preventing proliferation and at the same time preserving her own nuclear capabilities. When hypocrisy and prudence clash prudence is likely to prevail.

The second part of the criticism is much less acceptable. States acquire nuclear weapons for particular reasons usually associated with perceptions of their own security. What Britain does is unlikely to have any impact on the security problems facing South Africa, Israel, India, Pakistan or Brazil. In each of these cases it is the regional balance of power which is crucial. It is what potential rivals do which is all important. Britain's example is unlikely to have any great effect on the wider problem of proliferation. If further proliferation occurs, as seems likely, Britain faces the prospect of other nuclear powers in the world which may in the future threaten her interests. Britain's continuing possession of nuclear weapons provides added options which would not be available if she were a non-nuclear state.

If a convincing case can be made for the continuation of a British nuclear deterrent, the question arises whether Trident is the most cost-effective system available. One of the problems with Trident, as we have seen in Chapter 4, is that it is very expensive. A force which absorbs round 16 per cent of the new equipment budget is bound to create opportunity costs in other areas of the defence programme. There is certainly a case that when the decision was being made in 1979–80 much more serious consideration should have been given to a cheaper deterrent, even if it would have been less capable than Trident.[9] Trident is a 'Rolls-Royce deterrent' whereas Britain could have got away with a 'Ford Sierra'! The decision has been taken and provided there is no radical change in government policy (for economic or arms control reasons), the chances are that so much money will have been spent on the Trident programme by the time of the next election that the choice will be Trident or nothing. In such circumstances the arguments in favour of continuing with Trident seem quite sound. After all, it will only represent around 3 to 5 per cent of the defence budget over a twenty-year period which is certainly affordable given the unique strategic capability which it provides.

Two main questions arise from this judgement. Firstly, what role should Britain play, if any, in the strategic arms negotiations between the United States and the Soviet Union in the future? Secondly, should Britain continue its strategic nuclear relationship with the United States after Trident or plan to collaborate with France?

After the INF Agreement in December 1987, multilateral arms control is back on the tracks after the difficulties of the late 1970s and early 1980s. Indeed, for the first time in history, the Superpowers have reduced the number of nuclear weapons at the intermediate level and are seriously negotiating about cuts of up to 50 per cent in strategic nuclear forces. As a result the pressure is likely to build up on the British to include Trident in these negotiations. The pressures are likely to come from domestic constituencies as well as from the Soviet Union (and maybe from the United States). So far the British government's position has been that Britain would only include Trident in the discussions if 'substantial reductions' were made in Superpower arsenals. Whether 50 per cent reductions would qualify is far from clear. A START agreement along these lines would after all only bring the Superpowers down to the levels of the mid-1970s. Nevertheless, a START agreement would give a major political and psychological boost to multilateral arms control and to *détente* between East and West. In such circumstances there would be a strong case, following such an agreement, for Britain to associate herself more closely than she has in the past with further reductions. What form might this take?

When it is deployed Trident will represent roughly an eightfold increase in Britain's nuclear capabilities. No convincing argument has yet been put forward to justify such an enormous expansion of Britain's nuclear forces. Concern about improvements in Soviet defensive capabilities and the need for some cushion in case of developments in anti-submarine technology suggest the value of *some* extra capacity. An eightfold expansion, however, of a force which is predominantly concerned with counter-value targeting seems somewhat excessive. The government's case that the British force will still only represent a minimum deterrent is not very convincing. That maybe true of Polaris. It is not true of Trident, though, with its greater number of warheads, improved penetration and longer range. Some latitude, therefore, now seems to exist for Britain to make a contribution to the arms control process. There is a strong case to be made that getting back to the top table at which arms control negotiations take place would be in Britain's political and strategic interests and could help to continue the momentum of East–West dialogues.

The second question concerns the continuation of the nuclear relationship with the United States . In the past, whenever the question of Anglo-American or Anglo-French cooperation has been considered, Britain has always opted for the much greater advantages derived from a close nuclear partnership with the United States. Despite a certain loss of political and strategic independence, successive governments have come to the conclusion that the financial, technological and indeed strategic utility of the American connection outweighs the disadvantages of dependency. Whether this will be so in the future is a matter which will need to be considered very carefully.

For the moment the preference for a nuclear partnership with the United States is not surprising. Trident and the range of technological and intelligence assistance provided by the United States is far superior to anything which France could offer. Sustaining that relationship will probably remain in British interests but other pressures may seem likely to push Britain towards greater collaboration with France in the future. These pressures may include a declining US interest in assisting Britain in the nuclear field because of the growing importance of its strategic relationship with the Soviet Union or a gradual reorientation of US foreign and defence policy away from Europe. The Reykjavik summit, the INF Agreement and the *Discriminate Deterrence* report have all indicated the need for greater European defence cooperation. During the latter part of 1987 Britain and France discovered a mutual interest in opening discussions about future nuclear collaboration.[10] Although the traditional obstacles to close Anglo-French nuclear cooperation remain (like French independence and Britain's difficulty in passing on US information to a third party), both London and Paris are showing increased enthusiasm for exploring the possibility of associated projects.

The notion of keeping longer-term options open in the nuclear field would appear to be a sensible one. The British government is clearly aware of the danger that a growing nuclear relationship with France could undermine its long established and fruitful partnership with the United States. These two relationships, however, need not be mutually exclusive. It would make sense for Britain to try to encourage trilateral nuclear cooperation with the United States and France. Given the growing pragmatism in the French approach to defence and towards the United States in the 1980s, such collaboration may prove to be a practical possibility. Certainly economic and technological pressures are likely to make this more attractive for France (as well as for Britain) in the future. A policy of continuing cooperation with the United States while building the foundations of a nuclear relationship with France would appear to be very much in Britain's interests in the longer term.

One of the purposes of a defence review would be to consider this question of long-term direction in nuclear policy. The changing nature of Anglo-American and Anglo-French nuclear cooperation would clearly need to be kept under review, to ensure that conflicts of interest did not arise. The implications of arms control negotiations and agreement would also need to be constantly reassessed. All this adds to the need for a regular review process.

What the foregoing analysis suggests is that there is a good case for the continuation of the British nuclear deterrent and for the foreseeable future there is no realistic alternative to Trident. There would seem to be a case for including Trident in the next round of the START negotiations once the aim of 50 per cent reductions in Superpower arsenals has been achieved. This could involve a decision by the British government to cut the number of warheads on the Trident missiles. Above all, what this analysis indicates is that there probably is no room for substantial savings in the nuclear field in the near or even medium-term future. If savings are to be made they will have to be achieved in other areas.

HOME DEFENCE

In the 1981 defence review, the Ministry of Defence scrutinised home defence and concluded that 'we cannot reduce our effort in direct defence of the United Kingdom'.[11] The White Paper, *The Way Forward*, pointed out that Britain played a crucial role in support of NATO, as a key forward base for land and air forces from across the Atlantic and as the main base for her own efforts in reinforcing the Continent and conducting maritime tasks. Britain, it was argued, was therefore a primary target for the Soviet Union in the event of war. Given the increasing reach and quality of Soviet conventional forces the task of defending Britain was becoming more difficult.

The government decided that planned capabilities were already less than they should be and that 'more needed to be done'. In the field of air defence the commitment to both the Tornado (F2) and the Nimrod early-warning aircraft was reinforced and it was announced that the stocks of air-to-air and surface-to-air missiles would be improved.[12] More UK-based fighters were also to be made available.

At the same time, the government announced its determination to continue the build-up of a balanced mine counter-measures force for the defence of shipping lanes around Britain and to expand the Territorial Army for home defence (and for operations in Germany).

Since 1981 some of these improvement have been carried out but there is no doubt that UK home defence remains the weakest pillar of Britain's defence effort in the late 1980s. Britain remains a vital reinforcement base for NATO's forward military forces and this will continue to be the case whether the numbers of American troops in Europe remain the same as now or whether some return to the United States, as part of the need to cut defence costs in an attempt to lessen the large US budget deficit. In the latter case, the need to reinforce American ground and tactical air forces will be an even more important priority, and the need to provide effective UK home defence will correspondingly acquire greater significance. This would also necessitate more NATO infrastructure spending and the pre-positioning of American equipment not only on the Continent but in Britain as well.

The consensus amongst defence analysts remains that not enough attention is being given to the defence of the home base. It remains vital for the security of the state and for the defence of the NATO Alliance as a whole. Just as in 1981 more needs to be done in this area of defence policy, not less. There would appear, therefore, to be no substantial savings to be made in this field either. Taking account of the lessons emanating from the cancellation of the Nimrod early-warning aircraft, the review process would need to look carefully at long-term planning on home defence (in the air, at sea and on land) and concentrate on improving the provision for this vital, but hitherto neglected, area of defence capability.

THE DEFENCE OF THE CENTRAL FRONT

A fresh look at NATO strategy and Britain's contribution to it is certainly necessary at present if for no other reason than the INF Agreement of 1987. Since 1967 the Alliance has attempted to provide itself with a range of nuclear and conventional capabilities to fulfil the requirements of its Flexible Response strategy.[13] In recent years, there has been a recognition that the overemphasis on the early use of nuclear weapons not only created strains within the Alliance but also undermined the credibility of the deterrent strategy. As a result there has been an attempt to improve the flexibility of the strategy further by scaling down the dependence on nuclear weapons and improving the Alliance's conventional capabilities. This can be seen by the 1977 decision to increase Alliance spending by 3 per cent in real terms, the 1978 Long Term Defence Improvement Programme, and the 1983 Montebello decision to reduce the number of tactical nuclear weapons by 1400 warheads.

In one sense, the INF Agreement reinforces this trend towards less dependence on nuclear weapons, but it does create a very real difficulty for the strategy of Flexible Response. Taking away an important category of medium-range nuclear missiles leaves a significant gap in a strategy which emphasises the escalatory link between different levels of military capability from the conventional weapons, through tactical and theatre nuclear forces to the ultimate threat to use strategic nuclear forces. Consequently, although the INF Agreement has been supported because it makes good political sense in helping to sustain better East–West relations, it has created anxiety in Western Europe in particular because it undermines the logic of the existing strategy.[14]

The Alliance faces an acute dilemma over the INF Agreement as a result. NATO can respond by filling the gap with air-launched or sea-launched cruise missiles and thereby reinforce Flexible Response. Such 'compensating adjustments' would not break the letter of the INF Agreement, but might be perceived as breaking the spirit of it. This might well lead to new domestic discontent similar to the early 1980 demonstrations when cruise and Pershing II missiles were deployed. It would also anger the Soviet Union and thereby weaken the chances of further arms control agreements. Both domestic opposition groups and the Soviet leadership might argue that NATO was planning to take out one category of weapons with one hand while attempting cynically to replace them with another category of weapons with the other hand. The reaction might well be very hostile.

If the NATO Alliance does nothing about this gap in capabilities then the strategy of Flexible Response will be undermined. It is true that the strategy will not be wholly ineffective as a result of the withdrawal of cruise and Pershing II missiles because of the continuing existence of other nuclear capabilities. The credibility of the threat of step-by-step escalation, however, would clearly be weakened if the Alliance was unable to modernise its nuclear capabilities at the intermediate level.

Faced with this dilemma NATO can either face up to the consequences of making 'compensating readjustments' or learn to live with a strategy which has reduced credibility. Alternatively, the Alliance could replace Flexible Response with a new strategic concept. In recent years, Flexible Response has come in for a great deal of criticism not only from outside the defence establishment but also from a number of former high-ranking officers and civil servants. Members of the peace movement have been especially critical of NATO strategy, arguing that the notion of step-by-step escalation is very dangerous. They have not been alone in expressing this anxiety. A well-known American strategist Morton Halperin has argued: 'NATO doctrine is that we will fight with conventional weapons until we are losing, then we will fight with tactical nuclear weapons until we are losing, and then we will blow up the world'.[15] This concern with the dangers of nuclear escalation, and the problem which this creates for the credibility of the deterrent effect of Flexible Response, has been commented on by a range of individuals like Sir Frank Cooper, Lord Cameron and Lord Mountbatten who spent their careers at the highest level of the British defence establishment. Lord Mountbatten, who held the post of Chief of the Defence Staff from 1959 to 1965, was particularly worried about the role of battlefield nuclear weapons in NATO strategy. He pointed out that NATO believed that

> were hostilities to break out in Western Europe, such tactical nuclear weapons could be used in field warfare without triggering an all-out nuclear exchange leading to the final holocaust. I have never found this idea credible . . . As a military man I can see no use for any nuclear weapons which would not end in escalation, with consequences that no one can conceive.[16]

Much the same point was made by two other distinguished defence officials. Lord Cameron, who also held the post of Chief of the Defence Staff, and Sir Frank Cooper, who held the post of Permanent Under Secretary in the Ministry of Defence, argued in a pamphlet produced in 1984 that Flexible Response had 'ceased to retain much credibility as a deterrent, and needs to be replaced'.[17]

Defence planners in NATO have partially recognised the problem associated with the threat to use nuclear weapons early in a conflict. In recent years, there has been some attempt to emphasise conventional forces and de-emphasise nuclear capabilities. The difficulty has been that Flexible Response is such a vague concept that the commitment to conventional improvement has sometimes been offset by the continuing belief in some quarters that any attempt to scale down the nuclear capabilities of the Alliance will weaken deterrence. As a result of this ambivalence towards the role of nuclear weapons there has been an uncertainty and a lack of direction in NATO planning. The argument for a new strategic concept is that it will restore the sense of direction by pointing to the overriding importance of conventional improvements. What NATO needs is what might be called an 'Extended Firebreak Strategy' which will focus on the crucial import- ance of the threshold between conventional and nuclear war and move the Alliance towards a no-early-use posture.[18] In contrast to Flexible Response, nuclear weapons would perform the role of deterrents against nuclear attack but would cease to have an extensive war fighting role. Although there has been a recent trend towards improving conventional capabilities within the Alliance, a continuing ambivalence over whether to stress or to downgrade nuclear weapons has led to a loss of coherence in strategic and tactical planning. The notion of an 'Extended Firebreak' would help to overcome this ambivalence and provide an unambiguous encouragement for the task of improving the Alliance's conventional forces.

Critics of this approach would argue that such an idea is not practical politics because of the absence of the kind of political will necessary to fund the major improvements in conventional forces which would be necessary. This is a reasonable point to make but it reflects a continuing acceptance of the myth that there is a major disparity in conventional forces between East and West. Certainly there is an imbalance and in some areas Soviet superiority is worrying. Numerous studies have been produced recently including one by the WEU and one by the ex-US ambassador to the MBFR Talks, Jonathan Dean, which suggest that NATO is not as hopelessly outgunned as is often believed.[10] If the gap is not as great as it is usually portrayed then the chance of bridging it is not as difficult as many suggest. The adoption of a new strategy which de-emphasised nuclear weapons but which explained that there was a price to pay for such changes might well be acceptable to Western electorates. At the same time, pursuing vigorously an effective conventional arms reduc- tion agreement with the Soviet Union which produced balanced forces on both sides would make the costs less and the chances of political acceptance even greater.

Even if the idea of a new strategic concept is not acceptable clearly the case for conventional force improvements remains. So does the pursuit of a conventional arms control agreement with the Soviet

Union. In such a context Britain's contributions to the defence of the
Central Front would appear to be very important indeed, both from a
military and a political point of view. In military terms Britain and her
allies have a clear interest in keeping the nuclear threshold high. This
can only be done by sustaining conventional capabilities particularly
on the Central Front. If Britain pulled out or reduced its forces on the
Rhine it is difficult to see who would replace them. The political role
of BAOR is also of crucial importance. John Nott was right when he
argued:

> Nothing is more vital for British foreign policy than the interests of
> continental Europe. It dwarfs every other British interest ... The
> political contribution made to the cohesion of the alliance by the
> presence of our ground and air forces on the continent is out of all
> proportion to their military significance.[20]

The political contribution of British forces is very significant at a
time when the NATO Alliance is faced by important changes in East–
West relations and the anxieties which have been produced by
Gorbachev's peace offensive. Following the Reykjavik summit and
the INF Agreement, there was great concern in West Germany about
the coupling of American and European security. France has also
become anxious in recent years about neutralist tendencies in West
Germany. Given these uncertainties in the two major continental
European states, Britain has an important role to play in helping to
keep the Alliance together. The task of interpreting the European
view to the United States and the American view to the Europeans has
become increasingly important in recent years, and seems likely to be
of even greater significance in the future. Withdrawing part or all of
BAOR would not help in this process. It would send a signal to both
our European and American allies that Britain was less interested in
making a direct contribution to the most important area of Alliance
defence. The effect could well be to encourage Congress in the United
States to reduce American forces in Europe. If Britain does not see the
need to retain so many troops there why should the United States?
Keeping as large a contingent of American troops in Europe as
possible would seem to be in the interests of both Britain and Western
Europe and sustaining Britain's own commitment to the Central
Front contributes to this objective. Britain's continuing presence on
the Central Front, of course, does not guarantee a continuing
American commitment at the present level. All sorts of economic and

political pressures in the United States could result in troop cuts. Indeed, in the longer term if not sooner, this would seem to be very likely to occur. If this does happen Britain's contribution to European defence in both political and military terms will be even more significant than at present, in helping to fill the gap left by US troop withdrawals. A British contribution to greater European defence cooperation will in the first instance help to head off complaints from some in the United States that Europe is not pulling its weight in Alliance defence.

Should the pressures to cut troop levels become overwhelming, however, Britain's commitment to even greater European defence collaboration would become far more important. The dilemma at present is that the development of a European defence identity, especially one developed outside the NATO framework, will lead the United States to believe that Europe could stand on its own feet in defence terms and as a result encourage those who want American troop levels to be cut. On the other hand, if the US is going to withdraw some of its forces anyway at some point in the future, then perhaps Europe ought to be doing everything it can to establish an effective framework of European defence to cope with the psychological as well as physical gaps which this will create. The continuing presence of substantial forces on the Continent can help Britain to play a role in dealing with this dilemma. That role must be to encourage the process of greater and more effective European coopera- tion, if not directly within the context of NATO at least within the spirit of the Atlantic Alliance. Mrs Thatcher's criticisms of Franco- German defence cooperation in late 1987 and early 1988 and her warning of the danger of setting up substructures within the Alliance may suggest to some an oversensitivity to this problem.[21] The fact remains, however, that persuading the Americans that greater Euro- pean cooperation is designed to enhance the effectiveness of the European pillar within the Alliance, rather than to provide an alternative to the Atlantic partnership, is a very delicate task. Britain has an important dual objective. Firstly, to make sure that European defence arrangements do not in fact lead Europe towards an increas- ingly independent posture. And secondly, to convince the Americans that such independence is not intended in the cooperative arrange- ments which are taking place. Britain can only play such a role if its commitment to European defence is unquestioned.

A review of Britain's contribution to the defence of the Central Front, therefore, should lead to a continuation of Britain's commitment to stand by its Brussels Treaty obligations. At the same time, there would appear to be a need for a major review (not only in Britain but within the Alliance as a whole) of NATO's tactical and strategic doctrine in the light of the INF Treaty. In that review process Britain could play an important part in helping to move the Alliance even further away than at present from its overdependence on nuclear weapons and the existing ideas that nuclear weapons can be used as war-fighting instruments, which make no real sense. A new creative look at how NATO can defend itself conventionally, looking at the role of new emerging technology, deployment patterns, tactics, ammunition stocks, reserve forces and even considering *some* of the ideas of the alternative defence school, would appear to be worthwhile.[22] At present, NATO planners tend to be locked into the constraints imposed by the prevailing Flexible Response strategy. A new strategic concept would help release defence planners to produce their own 'new thinking' in the defence field.

THE DEFENCE OF THE EASTERN ATLANTIC

The defence of the Eastern Atlantic remains an important role for Britain. In 1981, however, it was decided to shift the emphasis away from surface ships towards maritime air power and submarines, thus implying some lowering of Britain's commitment to the defence of the Eastern Atlantic. The key question for a new defence review is whether the logic of the 1981 review still remains valid. As we have seen earlier, the impact of the Falklands War led to a renewed debate about the value of surface ships and a tendency by the government ever since to fudge the issue of priorities.

It is true that surface ships were important in the Falklands campaign and that the defence of the Eastern Atlantic does require a mix of forces. The Falklands War, however, reinforced the argument presented in the 1981 White Paper that surface ships are becoming increasingly vulnerable (sixteen of the twenty-three ships in the Falklands were hit). How much more vulnerable would such ships be when faced with the much more sophisticated Soviet air-launched and sea-launched missiles? Even if this were not the case, it must be accepted that the defence of the Falklands is not Britain's first priority and cannot be allowed to distort British defence policy. As the 1981 White Paper argued, Britain does need to retain 'a large and versatile ocean-going fleet'. The question is: what size fleet is required?

The choice facing defence planners is effectively between sustaining Britain's contribution to the Central Front or to the defence of the Eastern Atlantic. The argument presented in the last section suggests that nothing is more vital to Britain than the defence of the Continent itself. In contrast, important as the Eastern Atlantic undoubtedly is it would seem to be a lesser priority than the continental commitment. It is no good keeping the bridge to Europe from the United States open if the war has already been lost. A repeat of Second World War strategies in our technological age is hardly the ideal option. If NATO needs to specialise (and it does) there is a strong case for more US specialisation in the defence of the Eastern Atlantic. Although the 600 ship Navy has not been achieved and probably never will be, the build-up of US naval power in recent years has outstripped improvements in other areas. For the US the Atlantic is a crucial priority. It is conceivable that they might reduce their forces in Europe but it is highly unlikely that the US Navy would pull back in its Atlantic role. Britain must continue to provide support for the US Navy in the Eastern Atlantic but such support need not be of the same kind or on the same level as in the past.[23]

There is another question mark over the Navy's role in the Eastern Atlantic which is worthy of consideration. In the past year it has become increasingly clear that NATO has adopted a rather more offensive maritime strategy.[24] Forward defence has always been a part of NATO operations in order to defend the Atlantic routes (together with convoy protection and barrier defence), but more recently the Alliance has pursued a more offensive strategy based on American maritime doctrine. As the House of Commons Defence Committee heard in early 1988, British frigates and destroyers play an important part in this new strategy.[25] Forward defence, given the problems of defending the Atlantic against Soviet maritime power, is clearly important. There is a case, however, that the new offensive ideas are overly provocative, especially at a time when East–West relations are improving and increasing consideration is being given to more defensive ideas and confidence-building arrangements. NATO obviously cannot, and should not, bend over backwards in the direction of non-provocative gestures to the point where the defence of the Alliance is undermined. It is important, though, to understand that security is about political relationships as well as military balances. An acceptance of the need not to pursue overly provocative measures is something which is long overdue in both Western and Soviet strategy. Gorbachev's emphasis of common security ideas and the start of a debate in Soviet military circles about less offensive tactics should encourage the Western defence planners to begin their own 'new thinking'.[26] The questioning of the utility of the new, more offensive maritime strategy in the context of a broader interpretation of security would seem to be appropriate at the present time. This would need to be something which a new defence review might usefully consider.

The question mark over the future of the British surface fleet is not whether we have a fleet or not. The question is about relative priorities. Should Britain sustain the present commitment to a surface fleet of around fifty frigates and destroyers in service or accept a lower figure? Already there is a worry in naval quarters that the government's inability to place sufficient orders for new ships is likely to result in a decline in the surface fleet in the future. This policy of drift is not helpful to anyone. It would be far better for the government to conduct a review laying down the guidelines for British defence policy as a whole over the next five years into which its decisions about the future of the Royal Navy were explained and justified.

The conclusion of this analysis suggests that on balance the cuts which are clearly going to be necessary should, as in 1981, be found in reductions in the surface fleet. Even after taking into account all the arguments about the continuing value of surface ships in the defence of the Eastern Atlantic and their wider mobility and flexibility to deal with unforeseen events, the arguments in favour of the continental commitment appear stronger. The government has a choice. If it wishes to sustain the surface fleet at around the fifty ship level it could increase the resources devoted to defence. And there certainly would be a case for this given the flexibility which maritime power provides. If, however, defence spending remains more or less constant in real terms, once again something will have to go and analysis of relative priorities suggests the surface fleet should bear the brunt of the cuts.

OUT OF AREA CAPABILITIES

Cuts in the surface fleet would obviously have an effect not only on Britain's defence role in the Eastern Atlantic but on out-of-area roles as well. One of the most important lessons of the Falklands War was that token intervention forces are of no great value in an emergency. Had the Nott review been put into effect by the time of the Falklands campaign, the run down in amphibious capabilities and the very limited air and ground capabilities envisaged in *The Way Forward* for overseas operations would have meant that the task of retaking the islands would have been even more difficult than it was. Even though the cuts had not been implemented at the time of the war, it was a very close run thing. In the aftermath the government has made an attempt to upgrade out-of-area capabilities and seems to have accepted the need for 'viable' rather than 'token' intervention forces. The fact that Britain has specific responsibilities overseas in Gibraltar, Cyprus, Belize, the Falkland Islands and Hong Kong means that intervention capabilities cannot be allowed to fade away in a process of incremental reductions. Surface ships will also remain of value in certain specific circumstances in supporting wider British foreign policy objectives. This has been shown by the role played by British mine-sweeping operations in the Gulf in the latter half of 1987.

The main danger is that the Falklands War has encouraged a tendency in recent years for Britain to think in more expansive terms about playing a world role. The days when Britain was able to act as a world policeman are long over. The notion that Britain can fight 'brushfire wars around the world with the help of a deep-water navy' is illusory.[27] The incrementalist approach has increasingly led defence planners away from the sensible idea that Britain required able, but strictly limited intervention capabilities towards a much broader vision of Britain's international role. Britain's own interests and special expertise do suggest that she can make a useful contribution to help defend Western interests outside the geographically constrained NATO area, but this role should not be allowed to drain resources away from other more important capabilities. The notion that Britain can, and should, play a more important role on the world stage is corrosive. Such a role is vague and open-ended. It is not surprising that the naval lobby should emphasise such a role but the consequences of pursuing it would be to weaken Britain's defence effort in other more important areas. As John Nott has argued, such a role is unsupportable by rigorous analysis.[28]

The main thrust of the argument here is that although Britain should not allow its intervention potential to be eroded piecemeal as a result of constant cuts, there can be no open-ended commitment to a broader world role. As long as Britain retains overseas responsibilities and interests, a small, effective intervention force will be necessary but it must be remembered that this is a secondary priority.

The question of whether Britain can sustain its amphibious capabilities, as well as 'extra' surface ships to show the flag around the world, will need to be looked at carefully in the light of continuing responsibilities and the limited resources available. Sentimentality for traditional roles can have no place in the hard decisions about British defence policy for the 1990s and beyond. When each of the five main areas of defence policy are scrutinised this analysis suggests that (failing significant increases in defence spending) the key priorities should be the nuclear deterrent, home defence and the continental commitment. There is no doubt that the maritime contribution to Alliance defence and the continuing national requirement for out-of-area capabilities remain important. As such they must be funded *but at a level which does not undermine the three main pillars of British defence policy.*

There is, however, an important caveat to this judgement. British defence policy clearly has to strike a balance between national and Alliance requirements. Any review process cannot be made in isolation and should not be imposed on allies without detailed discussions. In deciding on its priorities Britain has to consider carefully the wider question of specialisation within NATO. It is clear from what has been said that the time is ripe for a major review of strategy within the Alliance. This would provide a useful occasion for British planners to try out their priorities and coordinate their long-term planning with that of their allies. It might be that as a result of this process our NATO partners would prefer Britain to downgrade its contribution to the Central Front and focus more on the Eastern Atlantic. If this were so Britain would still have to decide what was in her best interests but the preferences of our allies and the value of greater specialisation within the NATO Alliance would then provide an extra dimension to Britain's decisions about the future. Despite the value of vigorous analysis in determining defence priorities, such judgements are ultimately a matter of individual choice. This writer believes that such policies would be in the best interests of Britain but others will disagree. More important than the precise prescriptions, however, is the process by which decisions are arrived at. There is no doubt that the government can go on with its largely incrementalist approach. Indeed, some form of incrementalism, given the British policy-making system, is inevitable. The art is to try to combine the traditional elements of decision-making with an approach which encourages greater rigour and longer-term planning. If Britain is not to stumble along, weakening defence capabilities across the board and facing periodic defence reviews conducted in an atmosphere of crisis, a more systematic approach is going to be necessary. Those who argue for a major review are right but they are only partly right. What Britain needs is a regular process of review in which the language of priorities is spoken and an opportunity is created periodically for policy-makers to raise their heads from the minutiae of day-to-day problem-solving and provide long-term strategic direction, at present so clearly lacking in the conduct of British defence policy.

Appendix 1

Defence expenditure 1946–86 (Blue Book definition: calendar year)

Year	Defence expenditure, current prices £m	Implicit deflator	Defence expenditure, constant (1980) £m	As a percentage of GDP (at market prices) %
+* 1946	1575	5.9	26605	15.8
+* 1947	939	6.6	14120	8.8
* 1948	754	6.6	11344	6.4
* 1949	790	6.9	11363	6.4
1950	863	7.6	11302	6.7
1951	1315	8.5	15326	9.1
1952	1651	9.2	17757	10.5
1953	1719	9.5	18094	10.1
1954	1680	9.7	17177	9.4
1955	1567	10.1	15389	8.2
1956	1629	11.1	14605	7.9
1957	1590	11.7	13538	7.2
1958	1545	12.4	12440	6.8
1959	1571	12.8	12201	6.5
1960	1612	13.2	12127	6.3
1961	1732	13.7	12631	6.3
1962	1853	14.0	13176	6.4
1963	1905	14.4	13158	6.2
1964	2009	15.2	13163	6.0
1965	2127	16.1	13151	5.9
1966	2221	17.2	12878	5.8
1967	2427	17.7	13640	6.0
1968	2444	19.0	12846	5.6
1969	2292	20.1	11399	4.9
1970	2466	22.8	10770	4.8
1971	2763	25.7	10750	4.8
1972	3070	28.6	10712	4.8
1973	3479	31.9	10884	4.7
1974	4095	38.7	10563	4.9
1975	5177	48.2	10738	4.9
1976	6235	57.8	10778	4.9
1977	6863	64.3	10661	4.7
1978	7596	72.1	10530	4.5
1979	9006	83.9	10723	4.6
1980	11489	100.0	11489	5.0
1981	12672	110.4	11474	5.0

Year	Defence expenditure, current prices £m	Implicit deflator	Defence expenditure, constant (1980) £m	As a percentage of GDP (at market prices) %
1982	14500	123.1	11774	5.2
1983	15872	130.8	12131	5.3
1984	17234	138.4	12450	5.4
1985	18283	148.3	12325	5.2
1986	18628	157.9	11796	5.0
**1987/8	18782			
**1988/9	19215			
***1989/90	20120			
***1990/1	21180			
***1991/2	22090			

* With estimates
** Estimates from the 1987 and 1988 Statement on Defence Estimates
*** Projected estimates announced in November 1988
+ Excludes fixed capital formation
Source: The National Income Accounts Blue Books (1987 and earlier).

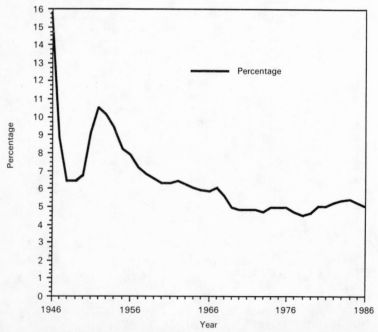

Figure A1.1 Defence expenditure as a percentage of GDP (1946–86)

Appendix 2

Defence expenditure compared with other items of government expenditure (at current prices)

Year	Defence	Education	NHS	Social security	Total government expenditure
+* 1946	1575			379	4635
+* 1947	939			494	4262
* 1948	754	309	249	541	4568
* 1949	790	346	427	597	5011
1950	863	370	478	614	4969
1951	1315	401	486	642	5724
1952	1651	444	497	753	6368
1953	1719	463	521	838	6710
1954	1680	502	537	844	6672
1955	1567	549	579	934	7088
1956	1629	636	633	998	7525
1957	1590	727	685	1046	7937
1958	1545	786	728	1271	8308
1959	1571	848	788	1367	8772
1960	1612	917	861	1402	9400
1961	1732	1012	930	1539	10307
1962	1853	1173	971	1645	11013
1963	1905	1281	1035	1884	11666
1964	2009	1417	1130	1981	12759
1965	2127	1585	1275	2284	14137
1966	2221	1700	1375	2442	14448
1967	2427	1893	1524	2747	16672
1968	2444	2096	1656	3170	18289
1969	2292	2250	1733	3388	19008
1970	2466	2532	1979	3709	20896
1971	2763	2899	2248	4068	23483
1972	3070	3407	2593	4833	26390
1973	3479	3995	2956	5224	30574
1974	4095	4649	3865	6450	39229
1975	5177	6626	5188	8353	51553
1976	6235	7318	6151	10611	58643
1977	6863	8341	6823	12693	61864
1976	6235	7318	6151	10611	58643
1977	6863	8341	6823	12693	61864
1978	7596	9153	7833	15629	72226
1979	9006	10310	9082	17794	85505
1980	11489	12753	11629	21394	104076

Year	Defence	Education	NHS	Social security	Total government expenditure
1981	12672	14314	13371	26564	116999
1982	14500	15278	14078	31063	128662
1983	15872	16340	15924	31123	138509
1984	17234	17061	16756	34515	146854
1985	18283	17420	17878	37587	157574
1986	18628	19521	19446	40616	162192
*1989/90	20120	19600	23200	51000	167100

+ Excludes fixed capital formation
* Estimates
Source: The National Income Accounts Blue Books (1987 and earlier)

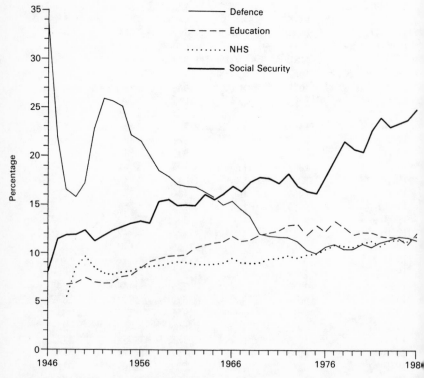

Figure A2.1 Defence expenditure compared with other items of public expenditure (1946–48).

Appendix 3

Defence expenditure as a percentage of total government expenditure

Year	Defence %	Education %	NHS %	Social security %
1946	33.9			8.1
1947	22.0			11.5
1948	16.5	6.7·	5.4	11.8
1949	15.7	6.9	8.5	11.9
1950	17.3	7.4	9.6	12.3
1951	22.9	7.0	8.4	11.2
1952	25.9	6.9	7.8	11.8
1953	25.6	6.9	7.7	12.4
1954	25.1	7.5	8.0	12.6
1955	22.1	7.7	8.1	13.1
1956	21.6	8.4	8.4	13.2
1957	20.0	9.1	8.6	13.1
1958	18.5	9.4	8.7	15.3
1959	17.9	9.6	8.9	15.5
1960	17.1	9.7	9.1	14.9
1961	16.8	9.8	9.0	14.9
1962	16.8	10.6	8.8	14.9
1963	16.3	10.9	8.8	16.1
1964	15.7	11.1	8.8	15.5
1965	15.0	11.2	9.0	16.1
1966	15.3	11.7	9.5	16.9
1967	14.5	11.3	9.1	16.4
1968	13.3	11.4	9.0	17.3
1969	12.0	11.8	9.1	17.8
1970	11.8	12.1	9.4	17.7
1971	11.7	12.3	9.5	17.3
1972	11.6	12.9	9.8	18.3
1973	11.3	13.0	9.6	17.0
1974	10.4	11.8	9.8	16.4
1975	10.0	12.8	10.0	16.2
1976	10.6	12.4	10.4	18.1
1977	11.0	13.4	11.0	20.5
1978	10.5	12.6	10.8	21.6
1979	10.5	12.0	10.6	20.8
1980	11.0	12.2	11.1	20.5
1981	10.8	12.2	11.4	22.7
1982	11.2	11.8	10.9	24.1
1983	11.4	11.7	11.4	23.1

Appendix 3 *cont.*

Year	Defence %	Education %	NHS %	Social security %
1984	11.7	11.6	11.4	23.5
1985	11.6	11.0	11.3	23.8
1986	11.4	12.0	11.9	25.0

Source: The National Income Accounts Blue Books (1987 and earlier).

Notes and References

Preface

1. C. J. Bartlett, *The Long Retreat* (London: Macmillan, 1972); L. W. Martin, *The Long Recessional*, Adelphi Paper No. 61 (London: IISS, 1969); C. Barnett, *The Collapse of British Power* London: Penguin, 1972); C. Coker, *A Nation in Retreat?* (London: Brassey's, 1986).
2. See R. Rosecrance, *Defense of the Realm: British Strategy in the Nuclear Epoch* (New York: Columbia University Press, 1968); D. Greenwood, 'Constraints and Choices in the Transformation of Britain's Defence Effort', *British Journal of International Studies*, vol. 2, no. 1, April 1976; P. Gore-Booth, *With Great Trust and Respect* (London: Constable, 1974); and A. Cyr, *British Foreign Policy and the Atlantic Area* (London: Macmillan, 1979).
3. Coker, *A Nation in Retreat*, pp. 1, 8, 11, 12.
4. Ibid, p. 11.
5. M. Edmonds, 'Planning Britain's Defence, 1945–85; Capability, Credibility and the Problem of Time', in M. Edmonds (ed.), *The Defence Equation* (London: Brassey's, 1986), p. 18.
6. *Statement on the Defence Estimates, 1986*, Vol. 1 (London: HMSO, 1986), Cmnd 9763-1 and *Statement on the Defence Estimates 1987*, Vol. 1 (London: HMSO, 1987), Cm. 101–1.

Introduction: The Incrementalist Approach to British Defence Policy

1. D. Braybrooke and C. E. Lindblom, *A Strategy of Decision: Policy Evaluation as a Social Process* (New York: The Free Press, 1963). See also C. E. Lindblom, *The Intelligence of Democracy: Decision-making Through Mutual Adjustment* (New York: The Free Press, 1965).
2. R. Rose, *What Is Governing? Purpose and Policy in Washington* (Englewood Cliffs, NJ: Prentice-Hall, 1978), p. 125.
3. Ibid.
4. Ibid, p. 126.
5. W. R. Schilling, 'The Politics of National Defense: Fiscal 1950', in W. R. Schilling, P. Y. Hammond and G. H. Snyder, *Strategy, Politics and Defense Budgets* (New York: Columbia University Press, 1962), p. 10.
6. Ibid, p. 14.
7. Ibid, pp. 12–13.
8. S. P. Huntington, *The Soldier and the State: The Theory and Politics of Civil–Military Relations* (Cambridge, Mass: Harvard University Press, 1957), p. 418.
9. See K. Booth, *Strategy and Ethnocentrism* (London: Croom Helm, 1979).

10. C. S. Gray, *Nuclear Strategy as a National Style* (London: Hamilton Press, 1986), p. 34.
11. L. D. Epstein, 'British Foreign Policy', in R. C. Macridis (ed.), *Foreign Policy in World Politics* (Englewood Cliffs, NJ: Prentice-Hall, 1962), p. 32.
12. 'Memorandum on the Present State of British Relations with France and Germany', by Eyre Crowe, 1 January, 1908, FO 371/257. See G. P. Gooch and H. Temperley (eds), *British Documents on the Origins of the War, 1898–1914*, Vol. III (London: HMSO, 1928), p. 397.
13. Quoted in Michael Howard, 'The British Way in Warfare: A Reappraisal', in M. Howard, *The Causes of War* (London: Allen & Unwin, 1984), p. 189.
14. Ibid.
15. 'Memorandum' by Eyre Crowe.
16. M. Wright (ed.), *Theory and Practice of the Balance of Power, 1486–1914* (London: Dent, 1975), pp. xvi–xvii.
17. F. S. Northedge, 'British Foreign Policy', in F. S. Northedge (ed.), *The Foreign Policy of the Powers* (London: Faber and Faber, 1968), pp. 150–85.
18. Ibid.
19. For a survey of the literature see C. Howard, *Britain and the Casus Belli, 1822–1920: A Study of Britain's International Position from Canning to Salisbury* (London: Athlone Press, 1974), p. 2.
20. Chamberlain, Memorandum, 29 March 1898, J. L. Garvin, *The Life of Joseph Chamberlain* (London: Macmillan, 1932–69), p. 260.
21. Howard, *The Causes of War*, p. 5.
22. Ibid, p. 172.
23. B. H. Liddell Hart, *The British Way in Warfare* (London: Faber & Faber, 1932), p. 7.
24. Ibid, pp. 25–6.
25. Ibid, pp. 35–7.
26. J. Corbett, *Some Principles of Maritime Strategy* (London: Conway Maritime Press, 1972). This book was first published in 1911. Corbett also published *Drake and the Tudor Navy* in 1897; *The Successors of Drake* in 1900; *England in the Seven Years War* in 1907; and *The Trafalgar Campaign* in 1910. Corbett was a lecturer at the Royal Naval College, Greenwich and historical adviser to the Admiralty. See also M. Howard, *The Causes of War*, p. 193.
27. Corbett, *Some Principles of Maritime Strategy*, p. 51.
28. Ibid.
29. Howard, *The Causes of War*, pp. 199–200.
30. Ibid, p. 200.
31. Ibid, p. 193.
32. Ibid.
33. Ibid, p. 200.
34. Ibid.
35. Hansard, 3, ccxxiv, col. 1099, 31 May 1875.
36. Armstrong, De Witt, *The Changing Strategy of British Bases* (unpublished Ph.D. thesis, Princeton University, 1960), p. 32.

37. J. C. Garnett, 'British Strategic Thought', in J. Baylis (ed.), *British Defence Policy in a Changing World* (London: Croom Helm, 1977), pp. 162–3.

38. J. Richardson (ed.), *Policy Styles in Western Europe* (London: Allen & Unwin, 1982), p. 2.

39. See J. E. S. Hayward, 'National Aptitudes for Planning in Britain, France and Italy', *Government and Opposition*, vol. 9, no. 4, 1974; 'Institutional Inertia and Political Impetus in France and Britain', *European Journal of Political Research*, vol. 4, no. 4, 1976; and J. E. S. Hayward and R. Berki, *State and Society in Contemporary Europe* (Oxford: Robertson, 1979). See also J. Richardson, A. G. Jordan and R. Kimber, 'Lobbying, Administrative Reform and Policy Styles', *Political Studies*, vol. xxvi, no. 1, 1978; J. Richardson and A. G. Jordan, *Governing Under Pressure: The Policy Process in Post-Parliamentary Democracy* (Oxford: Robertson, 1979); and A. G. Jordan and J. Richardson, 'The British Policy Style or the Logic of Negotiation', in Richardson (ed.), *Policy Styles in Western Europe*.

40. Jordan and Richardson, 'The British Policy Style or the Logic of Negotiation', p. 81.

41. Ibid.

42. Ibid, p. 85.

43. Ibid, p. 92.

44. M. Edmonds, 'The Higher Organization of Defence in Britain, 1945–85: The Federal–Unification Debate', in Edmonds (ed.), *The Defence Equation* (London: Brassey's, 1986), p. 57.

45. *Central Organization of Defence* (London: HMSO, 1946), Cmnd 6923.

46. Viscount Montgomery of Alamein, *The Memoirs of Field Marshal Montgomery* (London: World Publishing, 1958), pp. 427–46.

47. M. Edmonds, 'The Higher Organization of Defence in Britain, 1945–75', in Edmonds, *The Defence Equation*, p. 61.

48. Ibid, p. 62.

49. See W. P. Snyder, *The Politics of British Defence Policy, 1945–62* (Ohio: Ohio State University Press, 1964), pp. 123–204.

50. *Central Organization for Defence* (London: HMSO, 1958), Cmnd 476.

51. Edmonds, *The Defence Equation*, p. 67.

52. *Central Organization for Defence* (London: HMSO, 1963), Cmnd 2097.

53. Ibid, pp. 1–2.

54. Ibid, pp. 2–6.

55. Edmonds, *The Defence Equation*, p. 71.

56. Ibid, p. 72.

57. Ibid, p. 74.

58. *Statement of Defence Estimates, 1983*, Vol. 1 (London: HMSO, 1983), Cmnd 8951-1, p. 29.

59. 'Letters to the Editor', *The Times*, 23 March 1984.

60. Ibid.

61. Ibid.

62. It would appear that the Services have retained a great deal of influence over the formulation of policy. Confidential interviews November 1987.

1 The Continental Commitment Versus a Maritime Strategy

1. M. Howard, *The Continental Commitment: The Dilemma of British Defence Policy in the Era of Two World Wars* (London: Temple Smith, 1972) and Brian Bond, *British Military Policy Between the Two World Wars* (Oxford: Clarendon, 1980).

2. 'Defence Expenditure in Future Years: Interim Report by Minister for the Coordination of Defence', 15 December 1937. C.P. 316 (37). This represented something of a reversal of the Chiefs of Staff's arguments of 1934–5 that it was vital to hold Belgium to prevent an air attack on Britain and the political importance of land forces fighting by the side of France. See Howard, *The Continental Commitment*, p. 117.

3. 'European Appreciation' by the Chiefs of Staff Committee, 20 February 1939. DP (P) 44.

4. P.H.P. (44) 27 (O), (Final), 9 November 1944. CAB 81/95.

5. For a discussion of wartime planning see J. Baylis, 'British Wartime Thinking About a Post-war European Security Group', *Review of International Studies*, no. 9, 1983. See also J. Lewis, *Changing Direction: British Military Planning for Post-war Strategic Defence, 1942–47* (London: Sherwood Press, 1988).

6. P.M. to Eden, 25 November 1944, FO 371/40723, U8472/180/70.

7. See Sir Llewellyn Woodward, *British Foreign Policy in the Second World War*, Vol. 5 (London: HMSO, 1976), pp. 194–6.

8. The author has dealt in more detail with the debates about these alliances in 'Britain and the Dunkirk Treaty: The Origins of Nato', *The Journal of Strategic Studies*, vol. 5, June 1982 and 'Britain, the Brussels Pact and the Continental Commitment', *International Affairs*, vol. 60, 1984.

9. Bevin to Inverchapel, Telegram No. 1032, 26 January 1948, FO 371/73045, Z561/273/72G.

10. CAB 79, 54 COS (46) 187th Meeting.

11. See Viscount Montgomery of Alamein, *Memoirs of Field Marshal Montgomery*, pp. 498–500.

12. Ibid, p. 502.

13. E. Barker, *The British Between the Superpowers, 1945–50* (London: Macmillan, 1983), p. 155.

14. D. Greenwood, 'The 1974 Defence Review in Perspective', *Survival*, September–October, 1975.

15. *The United Kingdom Defence Programme: The Way Forward* (London: HMSO, 1981), Cmnd 8288.

16. Ibid, p.5.

17. Ibid, p. 6.

18. Ibid, pp. 8–10.

19. See Admiral of the Fleet Lord Hill-Norton, 'Return to a National Strategy', in J. Baylis (ed.), *Alternative Approaches to British Defence Policy* (London: Macmillan, 1983), pp. 117–18.

20. *The Times*, 17 August 1983.

21. *The Times*, 24 May 1984.

22. *The Times*, 19 May 1981.
23. See M. Chichester and J. Wilkinson, *The Uncertain Ally* (London: Gower, 1982), p. 56.
24. *Omega Report on Defence Policy* (Adam Smith Institute, September 1983).
25. Sir Henry Leach, 'British Maritime Forces: The Future', *RUSI Journal*, September 1982, p. 12.
26. Ibid.
27. Admiral of the Fleet Lord Hill-Norton, 'Return to a National Strategy', p. 117.
28. Ibid, p. 130.
29. See Field Marshal Lord Carver, 'Getting Defence Priorities Right', and Marshal of the Royal Air Force Lord Cameron, 'Alternative Strategies: Strategy, Tactics and New Technology', in Baylis (ed.), *Alternative Approaches to British Defence Policy*.
30. Field Marshal Lord Carver, ibid, p. 83
31. Ibid.
32. Ibid, p. 76.

2 European Versus Global Defence

1. See in particular A. Verrier, *Through the Looking Glass: British Foreign Policy in the Age of Illusions* (London: Jonathan Cape, 1983) and D. Dilks (ed.), *Retreat from Power: Studies in Britain's Foreign Policy of the Twentieth Century* (London: Macmillan, 1981).
2. Numerous studies on British foreign policy quote Dean Acherson's famous remark that 'Britain has lost an Empire and not yet found a role'.
3. See A. Pierre, *Nuclear Politics: The British Experience with an Independent Strategic Force, 1939–70* (London: Oxford University Press, 1972), p. 316.
4. See P.H.P. (44) 27 (O), (Final, 9 November 1944, CAB 81/95; Sir Llewellyn Woodward, *British Foreign Policy in the Second World War* (London: HMSO, 1976); and J. Lewis, *Changing Direction: British Military Planning for Post-war Strategic Defence, 1942–47* (London: Sherwood Press, 1988).
5. Ibid.
6. This was a point emphasised by Sir Harold Macmillan in an interview with the author on 28 August 1979.
7. This was clear from Britain's participation in the Dunkirk Treaty, the Brussels Pact and NATO.
8. See E. Barker, *The British Between the Superpowers, 1945–50* (London: Macmillan, 1983), pp. 48–52 and R. A. Best, Jr, *'Co-operation with Like-Minded People's': British Influences on American Security Policy, 1945–49* (New York: Greenwood Press, 1986).
9. Barker, *The British Between the Superpowers, 1949–50*, p. 196.

10. P. Darby, *British Defence Policy East of Suez, 1947–68* (London: Oxford University Press, 1973), pp. 10–31.
11. Ibid, p. 116.
12. R. Rosecrance, *Defense of the Realm: British Strategy in the Nuclear Epoch* (New York: Columbia University Press, 1968), pp. 263–4.
13. Darby, *British Defence Policy East of Suez, 1947–68*, pp. 252–5.
14. Ibid, chapter 5.
15. Ibid, pp. 244–8.
16. Ibid, p. 224.
17. Ibid, pp. 284–5.
18. Ibid, pp. 309–34.
19. In particular the original 'special capability' has given way to a 'general capability'.
20. *The United Kingdom Defence Programme: The Way Forward* (London: HMSO, 1981), Cmnd 8288, p. 11.
21. Ibid.
22. J. Wyllie, *The Influence of British Arms: An Analysis of British Military Intervention Since 1956* (London: Allen & Unwin 1984), p. 106.
23. *The Falklands Campaign: The Lessons* (London: HMSO, 1982), p. 31.
24. Ibid.
25. Ibid, p. 32.
26. See M. Chichester and J. Wilkinson, *The Uncertain Ally* (London: Gower, 1982) and *British Defence: A Blueprint for Reform* (London: Brassey's, 1987).
27. Ibid. See also Admiral of the Fleet Lord Hill-Norton, 'Return to a National Strategy', in Baylis (ed.), *Alternative Approaches to British Defence Policy*.
28. *The Falklands Campaign*, p. 32.

3   Europe Versus the 'Special Relationship'

1. In fact as Chapter 1 has shown the formal commitment to send reinforcements to the Continent in the event of war was not made until 23 March 1950.
2. See in particular M. Gowing, *Independence and Deterrence: Britain and Atomic Energy, 1945–52*, Vol. 1 (London: Macmillan, 1974); A. Bullock, *Ernest Bevin* (London: Heineman, 1983); and R. Edmonds, *Setting the Mould: The United States and Britain, 1945–50* (Oxford: Clarendon Press, 1986).
3. R. A. Best, Jr, *'Co-operation with Like-Minded People's': British Influences on American Security Policy, 1945–49* (New York: Greenwood Press, 1986).
4. E. Barker, *The British Between the Superpowers, 1949–50* (London: Macmillan, 1983), p. 127.
5. See P.H.P. (44) 27 (0), (Final), 9 November 1944.
6. Quoted in R. Crossman, *Diaries of a Cabinet Minister* Vol. I (London: Hamish Hamilton and Jonathan Cape, 1976), p. 95.
7. Gowing, *Independence and Deterrence*, p. 242.

8.   Ibid, p. 331.
9.   For a discussion of Anglo-American nuclear cooperation see T. J. Botti, The Long Wait: The Forging of the Anglo-American Nuclear Alliance (Connecticut: Greenwood Press, 1987); J. Simpson, *The Independent Nuclear State: The United States, Britain and the Military Atom* (London: Macmillan, 1983); and J. Baylis, *Anglo-American Defence Relations, 1939–84: The Special Relationship* (London: Macmillan, 1984).
10. H. Wilson, *The Labour Government, 1964–70* (London: Penguin, 1974), p. 80.
11. H. A. Kissinger, *The White House Years* (London: Weidenfeld & Nicolson, 1979), pp. 933–4.
12. See *The Daily Telegraph,* 19 December 1970 and 21 December 1970.
13. See Baylis, *Anglo-American Defence Relations, 1939–84*, p. 172.
14. For a discussion of the 'Committee of Four' and the Duff and Mason working parties see 'Planning for a Future Deterrent', in *The Times*, 4 December 1979.
15. As a result of the cuts in defence expenditure in Britain in 1975–6 the Chairman of the American Joint Chiefs of Staff, General Brown, described Britain's armed forces as 'pathetic' in 1976. He declared that 'They're no longer a world power, all they've got are generals, admirals and bands', *The Times*, 19 October 1976.
16. See *The Sunday Times,* 23 December 1979 and *The Observer,* 16 December 1979.
17. For an interesting discussion of these agreements see C. McInnes, *Trident: The Only Option?* (London: Brassey's, 1986).
18. See *The United Kingdom Trident Programme*, Defence Open Government Document 82/1, March 1982.
19. See *The Guardian,* 14 November 1979 and *The Times,* 11 December 1979.
20. Especially the Netherlands and Belgium.
21. See Baylis, *Anglo-American Defence Relations, 1939–84*, pp. 187–9, and D. Dimbleby and D. Reynolds, *An Ocean Apart* (London: Hodder, 1988).
22. See *The Observer,* 20 April 1986.
23. See 'Thatcher Backs Allotted Areas for Navies', in *The Times* 23 September 1987.
24. T. Taylor, 'Britain's Response to the Strategic Defence Initiative', *International Affairs*, vol. 62, 1986.
25. The text of the speech is contained in the *RUSI Journal*, vol. 130, March 1985, pp. 3–8.
26. Confidential interviews, November 1987.
27. *The Arms Control Reporter*, vol. 6:12 (Maryland: The Institute for Defence and Disarmament Studies, 1987), p. 403.
28. Confidential interviews, November 1987.
29. *Statement on Defence Estimates*, Vol. I (London: HMSO, 1985), Cmnd 9430-1, p. 18.
30. See *The Sunday Times,* 22 December 1985. See also L. Freedman, 'The Case of Westland and the Bias to Europe', *International Affairs*, vol. 63, 1986–7.

31.   See *The Sunday Times*, 20 December 1987.
32.   See *The Independent*, 29 January 1988.
33.   Ibid.
34.   *The Independent*, 30 January 1988.

4 Nuclear Weapons Versus Conventional Forces

1.   See M. Gowing, *Britain and Atomic Energy, 1939–45* (London: Macmillan, 1964).
2.   M. Gowing, *Independence and Deterrence: Britain and Atomic Energy, 1945–52*, Vol. 1 (London: Macmillan, 1974), pp. 160–93.
3.   See I. Clark and N. Wheeler, *The British Origins of Nuclear Strategy, 1945–55* (London: Oxford University Press, 1989).
4.   Gowing, *Independence and Deterrence*, pp. 224–7, 229–31, 233–4.
5.   Ibid.
6.   Ibid, pp. 440–3. See also R. Rosecrance, *Defense of the Realm: British Strategy in the Nuclear Epoch* (New York: Columbia University Press, 1968), p. 159.
7.   N. Wheeler, 'British Nuclear Weapons and Anglo-American Relations, 1945–54', *International Affairs*, vol. 62, 1985–6.
8.   Rosecrance, *Defense of the Realm*, p. 225.
9.   See P. Darby, *British Defence Policy East of Suez, 1947–68* (London: Oxford University Press, 1973), pp. 168–72.
10.   Ibid, pp. 244–82. It would be wrong, of course to argue that the priority given to the nuclear deterrent (exemplified by the 1963 Polaris Agreement) was solely responsible for the inadequacies of Britain's conventional forces. The Polaris programme, after all, only represented 2–3 per cent of the defence budget. To this, however, must be added the expenditure on the V-Bomber programme which continued, together with all the hidden costs associated with the nuclear programme which are not included in the accounting figures for nuclear weapons published in the annual White Papers. New nuclear weapons, for example, often represent a high proportion of the *new* equipment budget. Admirals are fond of complaining that expenditure on Polaris, in terms of running costs, comes out of the naval budget which creates opportunity costs for the Navy in terms of conventional naval forces. Also, with priority being given to nuclear forces it is the conventional forces which suffer during times of defence retrenchment. This was true in the mid-1960s just as it was in the mid-to-late 1970s and early 1980s.
11.   For a discussion of 'the long defence review' from 1964 to 1968 see Darby, *British Defence Policy East of Suez*, chapters 8 and 9.
12.   See L. Freedman, *Britain and Nuclear Weapons* (London: Macmillan, 1980), pp. 41–51.
13.   See D. Greenwood, 'The 1974 Review in Perspective', *Survival*, September–October, 1975.
14.   The ten areas included: Readiness; Reinforcement; Reserve Mobilisation; Maritime Posture; Air Defence; Communications, Command

and Control; Electronic Warfare; Rationalisation; Logistics; and Theatre Nuclear Modernisation. See *Statement on the Defence Estimates, 1979* (London: HMSO, 1979), Cmnd 7474.

15. See *The Times*, 4 December 1979.
16. *The United Kingdom Defence Programme: The Way Forward* (London: HMSO, 1981), Cmnd 8288.
17. Ibid, p. 5.
18. See *The Falklands Campaign: The Lessons* (London: HMSO, 1982), Cmnd 8758.
19. See J. Baylis (ed.), *Alternative Approaches to British Defence Policy* (London: Macmillan, 1983).
20. Lord Carver, 'Getting Defence Priorities Right' in Baylis, *Alternative Approaches to British Defence Policy*, pp. 80, 89.
21. Ibid, p. 80.
22. Ibid, p. 89.
23. Quoted in G. Segal *et al.*, *Nuclear War and Nuclear Peace* (London: Macmillan, 1983), p. 125.
24. There is no evidence that he expressed those doubts while he was Chief of the Defence Staff. Indeed, it would seem that he did not. Confidential interviews, November 1987 and correpondence with Lord Carver.
25. Michael Foot and Neil Kinnock.
26. There was a great deal of ambiguity in the 1964 election over what the Labour Party meant by their promise 'to renegotiate the Nassau deal'. See the different interpretations put forward in the *New Statesman*, 13 December 1963; *The Observer*, 17 November 1963; and *The Times* 13 April 1964. See also J. Baylis, *Anglo-American Defence Relations, 1939–84: The Special Relationship* (London: Macmillan, 1984), pp. 141–5.
27. *The Times*, 4 December 1979.
28. In the 1987 party conference there were warnings from the party's left wing (especially Mr Livingstone) not to abandon its unilateralist policies. See *The Guardian*, 2 October 1987.

5 Alliance Commitments Versus National Independence

1. As the Introduction has tried to show Britain has traditionally been somewhat ambivalent about alliances. British governments have had a 'reluctance to enter binding commitments' but this doesn't mean that Britain has not recognised the value of alliances at certain times.
2. See *The Memoirs of Lord Gladwyn* (London: Weidenfeld & Nicolson, 1972).
3. P.H.P. (44) 27 (0) (Final), 9 November 1944. CAB 81/95. See also Sir Llewellyn Woodward, *British Foreign Policy in the Second World War*, Vol. 5 (London: HMSO, 1976).
4. This is not, of course, a view which will be shared by everyone. Even if one accepts that the West must share some responsibility for the outbreak of the Cold War it is not surprising that Western politicians

should have pursued policies designed to contain the Soviet Union in the 1940s and 1950s given Soviet actions and pronouncements. Neither is it surprising that British politicians faced with the difficult task of withdrawing from Empire should have sought to reinforce her residual global interests through the wider security provided by various alliance arrangements. It is the author's view that despite the drawbacks of alliance memberships there were important advantages, both diplomatically and militarily, through Britain's participation in a number of collective security arrangements. Some were more useful than others. NATO has clearly been of much greater value than either SEATO or CENTO.

5. See R. Rose, *The Relation of Socialist Principles to British Labour Foreign Policy, 1945–51* (unpublished Ph.D. thesis, Oxford, 1959), pp. 333–4.

6. Harold Macmillan confirmed to the author that American financial pressure had been decisive in the Cabinet's decision to call off the military campaign. Interview 28 August 1979.

7. *Defence: Outline of Future Policy: 1957* (London: HMSO, 1957).

8. Exchange of notes constituting an agreement between the United States of America and the United Kingdon of Great Britain and Northern Ireland relating to intermediate range ballistic missiles. Signed at Washington, 22 February 1958', *United Nations Treaties Series*, 1958, vol. 307, no. 4451, pp. 208–14.

9. 'Polaris sales agreement between the Government of the United States of America and the Government of the United Kingdom of Great Britain and Northern Ireland. Signed at Washington, on 6 April 1963', *United Nations Treaty Series*, 1963, vol. 474, no. 6871, pp. 50–68.

10. All these agreements can be found in J. Baylis, *Anglo-American Defence Relations, 1939–84: The Special Relationship* (London: Macmillan, 1984).

11. How long it would take before the British force was seriously threatened through a cut-off of US assistance is classified. Estimates suggest sometime between eighteen and twenty-four months. Such estimates, however, are purely guesswork. Given the range of variables involved much would depend on the form the cut-off took and how 'seriously threatened' was defined.

12. Andrew Pierre has argued that as a result of the very close nuclear partnership Britain has 'undoubtedly lost a measure of her strategic independence'. A. Pierre, *Nuclear Politics: The British Experience with an Independent Strategic Force 1939–70* (London: Oxford University Press, 1972), p. 316.

13. Britain had severe reservations about the Lebanon operation. Confidential interviews, November 1987.

14. *The Observer*, 20 April 1986.

15. See *The Times*, 23 September 1987.

16. In the case of the Gulf operation, despite the attempt publicly to keep some distance from the US there does seem to have been some coordination of effort in practice. Ibid.

17. See L. Freedman, 'British Nuclear Targeting', *Defence Analysis*, vol. 1, no. 2, 1985, pp. 81–99.
18. See N. Wheeler, 'British Nuclear Weapons and Anglo-American Relations, 1945–54', *International Affairs*, vol. 62, winter 1985–6.
19. Hansard, vol. 537, col. 1897, 1 March (1955). As Lawrence Freedman notes Harold Macmillan, then Defence Minister, made a similar point in the same debate about the need to have 'influence over the selection of targets and the use of our vital striking forces'. Col. 2182, 2 March (1955).
20. M. Gowing, *Britain and Atomic Energy, 1939–45* (London: Macmillan, 1964), pp. 413–17.
21. S. Duke, *US Defence Bases in the United Kingdom* (London: Macmillan, 1987). See also J. Baylis, 'American Bases in Britain: The 'Truman–Attlee Understandings', *The World Today*, vol. 42, nos 8–9, August–September 1986.
22. *The Times*, 14 November 1958.
23. *Report on Defence: Britain's Contribution to Peace and Security* (London: HMSO, 1958), Cmnd 373.
24. Freedman, 'British Nuclear Targeting', p. 85.
25. Ibid.
26. Ibid, p. 90.
27. See also D. Ball, *Targeting for Strategic Deterrence*, Adelphi Paper No. 185 (London: IISS, 1983) p. 16 and G. Kemp, *Nuclear Forces for Medium Powers: Part One, Targets and Weapons Systems*: Parts II and III, *Strategic Requirements and Options*, Adelphi Papers Nos 106 and 107 (London: IISS, 1974).
28. Freedman, 'British Nuclear Targeting', p. 93.
29. J. Alford and P. Nailor, *The Future of Britain's Deterrent Force*, Adelphi Paper No. 156 (London: IISS, 1980), pp. 27–8.
30. Confidential interviews, November 1987. See also Freedman, 'British Nuclear Targeting', p. 94.
31. *The Future United Kingdom Strategic Nuclear Deterrent Force*, Defence Government Document, 80/23 (London: MOD, 1980), pp. 5–6.
32. Freedman, 'British Nuclear Targeting'.
33. Ibid, pp. 95–6.
34. See Lord Hill-Norton, 'Return to a National Strategy', in J. Baylis (ed.), *Alternative Approaches to British Defence Policy*; M. Chichester and J. Wilkinson, *The Uncertain Ally* (London: Gower, 1982) and *British Defence: A Blueprint for Reform* (London: Brassey's, 1987).
35. Lord Hill-Norton, 'Return to a National Strategy', p. 124.
36. Ibid.
37. Ibid.
38. Ibid, p. 136.
39. Ibid, p. 137.
40. For a discussion of the views of these members of 'the naval lobby' see C. Coker, *A Nation in Retreat?* (London: Brassey's, 1986), pp. 78–9.
41. The government's dilemma has been clear in the Defence White Papers which have been produced since 1982.

42.   Improvements have been made in Britain's intervention capabilities
       and the government maintained a force of 'around 50 surface ships'
       during the 1980s.

6   The Economic Challenge

1.   C. F. Bastable, *Public Finance* (London: Macmillan, 1895) quoted by
      W. P. Snyder, *The Politics of British Defence Policy, 1945–62* (Colum-
      bus: Ohio State University Press, 1964), pp. 181–2.
2.   See A. T. Peacock and J. Wiseman, *The Growth of Public Expenditure
      in the United Kingdom* (Princeton: Princeton University Press, 1961),
      pp. 168–9.
3.   Snyder, *The Politics of British Defence Policy, 1945–62*, p. 183.
4.   Ibid.
5.   *Statement Relating to Defence: 1948* (London: HMSO, 1948), Cmnd
      7327.
6.   D. Greenwood, 'Constraints and Choices in the Transformation of
      Britain's Defence Effort Since 1945', *British Journal of International
      Studies*, vol. 2, no. 1, April 1976.
7.   R. Rosecrance, *Defense of the Realm* (New York: Columbia University
      Press, 1968), p. 31.
8.   See Greenwood, 'Constraints and Choices', p. 19.
9.   See J. C. R. Dow, *The Management of the British Economy, 1945–60*
      (Cambridge: Cambridge University Press, 1964), p. 57.
10.   *Statement on Defence: 1952* (London: HMSO, 1952), Cmnd 8475.
11.   J. Slessor, *The Great Deterrent* (New York: Praeger, 1957), p. 126.
12.   Greenwood, 'Constraints and Choices', p. 24.
13.   Ibid.
14.   *Statement on the Defence Estimates, 1975* (London: HMSO, 1975).
15.   See *The Times*, 19 October 1976,
16.   *The Way Forward* (London: HMSO, 1981) Cmnd 8288.
17.   Greenwood, 'Constraints and Choices', p. 24.
18.   This question is discussed in J. Baylis, "Greenwoodery" and British
      Defence Policy', *International Affairs*, vol. 62, 1986.
19.   *The Government's Expenditure Plans 1986/7 to 1988/9*. See also M.
      Chalmers, *Trends in UK Defence Spending in the 1980s*, Peace Research
      Report No. 11, University of Bradford, September 1986. In November
      1988 the government announced revisions in defence spending plans
      for the period 1989 to 1992. In 1989/90 defence spending would fall by
      0.7 per cent in real terms. This was the third year that defence spending
      had fallen in real terms. Subsequent increases were dependent on an
      inflation rate of 3 to 3.5 per cent. (In 1988 the inflation rate was almost
      6 per cent.)
20.   D. Greenwood, 'Defence', in P. Cockles (ed.), *Public Expenditure
      Policy, 1985/6* (London: Macmillan, 1985).
21.   Ibid, pp. 117–18.
22.   See Chalmers, *Trends in UK Defence Spending in the 1980s*, p. 4.
23.   *The Sunday Times*, 12 January 1986.
24.   *The Third Report from the Defence Committee, Session 1984/5: Defence*

Commitments and Resources and the Defence Estimates 1985/6, Vol. 1, Part Two (London: HMSO, 1985).

25. Ibid.
26. See 'Defence Cuts Go Across the Board', *The Guardian*, 1 July 1986.
27. This will be a period when Trident expenditure will reach its peak and the government will also be trying to find resources for other expensive defence projects such as the Harrier GR5, the EFA, Type 23 frigates, AMRAAM, Sea Eagle, Stingray, Challenger tanks, Warrier, MRLS, the EH 101, and a new heavyweight torpedo.

7 The Technological Challenge

1. J. C. Garnett, 'Technology and Strategy', in J. Baylis, K. Booth, J. Garnett, P. Williams, *Contemporary Strategy: Theories and Concepts*, Vol. I (New York: Holmes & Meier, 1987), p. 95.
2. *Statement on the Defence Estimates, 1984*, Vol. I (London: HMSO, 1984), Cmnd 9227-1.
3. Ibid.
4. Ibid.
5. C. Coker, *A Nation in Retreat?* (London: Brassey's 1986), pp. 10–11.
6. See M. Chichester and J. Wilkinson, *British Defence* (London: Brassey's, 1987), p. 129. See also *The Independent*, 29 March 1988.
7. It was reported in 1988 that up to £4 billion of defence spending may have been soaked up by unforseen increases in equipment costs in general. See *The Independent*, 11 March 1988.
8. Chichester and Wilkinson, *British Defence*, pp. 122–3, and 129.
9. R. Rosecrance, *Defense of the Realm* (New York: Columbia University Press, 1986), p. 179.
10. 537 H.C. Deb. 2070–72.
11. Defe 4/81, COS(55) 104th meeting, Minute 2, 13 December 1955.
12. See J. E. Stromseth, *The Origins of Flexible Response* (London: Macmillan, 1988).
13. See *Diminishing the Nuclear Threat: NATO's Defence and New Technology* (London: The British Atlantic Committee, 1984).
14. Ibid.
15. Garnett, 'Technology and Strategy', pp. 103–9.
16. Ibid, p. 105.
17. *The Way Forward* (London: HMSO, 1981) Cmnd 8288, p. 4.
18. Ibid.
19. Garnett, 'Technology and Strategy', pp. 100–3.
20. Concern about Soviet ABM development led to the British Chevaline programme.
21. A. Pierre, *Nuclear Politics: The British Experience with an Independent Strategic Force* (London: Oxford University Press, 1972), pp. 201–7.
22. See *The Times*, 29 October 1987.
23. Ibid.
24. Ibid.
25. See *The Times*, 23 October 1987.
26. See W. Wallace, *The Foreign Policy Process in Britain* (London: Allen & Unwin, 1976), pp. 140–55.

8   The Political Challenge

1.   H. Laski, *Parliamentary Government: A Commentary* (London: Allen & Unwin, 1938), pp. 277–8.
2.   See D. Greenwood, 'Why Fewer Resources for Defence? Economics, Priorities and Threats', *The Royal Air Forces Quarterly*, vol. 14, no. 4, 1974, p. 278.
3.   Laski, *Parliamentary Government: A Commentary*.
4.   The exceptions occurred in the early 1950s with the Korean War and the late 1950s and early 1960s when nuclear weapons became a political issue.
5.   See P. M. Jones, 'British Defence Policy: The Breakdown of Interparty Consensus', *Review of International Studies*, vol. 13, no. 2, April 1987.
6.   See The Labour Party Manifesto, *Britain Will Win With Labour* (London, 1974).
7.   Jones, 'British Defence Policy', pp. 113–14.
8.   See The Labour Party Manifesto, *The New Hope for Britain* (London, 1983).
9.   See *The Conservative Party Manifesto, 1983* (London, 1983) and the SDP/Liberal Alliance Manifesto, *Working Together For Britain* (London, 1983).
10. Ibid.
11. *Labour Party Annual Conference Report, 1983* (London, 1984), pp. 160–1. Quoted by Jones, 'British Defence Policy', p. 120.
12. Labour Party NEC Statement, *Defence and Security for Britain* (London, 1984).
13. See J. Baylis and D. Balsom, 'Public Opinion and the Parties' Defence Policies', *The Political Quarterly*, vol. 57, no. 2, April–June 1986.
14. SDP Policy Statement No. 9, *Defence and Disarmament: Peace and Security* (London, 1985).
15. Ibid, p. 15.
16. For a discussion of the Alliance debate see Jones, 'British Defence Policy', p. 122.
17. Report of the Joint Liberal/SDP Commission, *Defence and Disarmament* (London, 1986).
18. For a discussion of the conference and the amendment calling for a non-nuclear defence policy see *The Guardian*, 24 September 1986.
19. H. C. Debs, vol. 100, no. 139, cols 719 and 727.
20. See *The Guardian*, 2 October 1987.
21. Ibid.
22. Mr Davies announced his resignation in a dramatic and bitter telephone call to the Press Association in the middle of the night on 14 June. He said: 'I am fed up with being humiliated by Neil Kinnock. He never consults me on anything. He goes on television and talks about defence, but he never talks to his defence spokesman ... I certainly did not agree with what he said on television. I cannot get up in the Commons and make a speech, when Mr Kinnock says all sorts of

things all over the place, one thing one day, and another thing the next. And he is supposed to be a future Prime Minister.' Defence also became an important issue in the leadership contest in the party in the summer of 1988 with Mr Benn and Mr Heffer criticising the changes in policy made by Mr Kinnock and Mr Hattersley.

23. See *The Independent*, 21 June 1988.
24. See *The Sunday Times*, 29 May 1988.
25. See *The Guardian*, 20 October 1987.
26. See *The Sunday Times*, 21 February 1988.
27. See *The Independent*, 4 March 1988.
28. See *The Sunday Times*, 21 February 1988.
29. See D. Calleo, *Beyond American Hegemony* (Oxford: The Alden Press, 1987).

9 Incrementalism Versus a Radical View

1. D. Greenwood, 'The Defense Policy of the United Kingdom', in D. J. Murray and P. R. Viotti, *The Defense Policies of Nations: A Comparative Study* (Baltimore: Johns Hopkins University Press, 1982), pp. 197–8.
2. Ibid, p. 201.
3. See *The Third Report from the Defence Committee, Defence Commitments and Resources and the Defence Estimates, 1985–6*, Vol. 1 (London: HMSO, 1985), and D. Greenwood, 'Defence', in P. Cockles (ed.), *Public Expenditure Policy, 1985/6* (London: Macmillan, 1985), pp. 117–18.
4. *The Third Report*, Vol. 1, ibid, p. xv.
5. Ibid.
6. *The Third Report*, Vol. 2, pp. 17–34.
7. *The Sunday Times*, 12 January 1986.
8. See *The Times*, 5, 6 October 1987.
9. *The Times*, 6 October 1987.
10. This was an argument put to the author in a series of confidential interviews in the Ministry of Defence in November 1987.
11. *The Way Forward* (London: HMSO, 1981) Cmnd 8288, p. 8.
12. See Lord Hill-Norton, 'Return to a National Strategy', in J. Baylis, *Alternative Approaches to British Defence Policy*, and K. Speed, *Seachange: The Battle for the Falklands and the Future of Britain's Navy* (London: Ashgrove Press, 1982).
13. This was another criticism raised in confidential interviews in the MOD, in November 1987.
14. For example, M. Chichester and J. Wilkinson, *The Uncertain Ally* (London: Gower, 1982).
15. Particularly compared with the Dutch and French experiences.
16. S. P. Huntington, *The Soldier and the State: The Theory and Politics of Civil–Military Relations* (Cambridge, Mass: Harvard University Press, 1959), p. 41.

17.　*The Times*, 6 October 1987.

18.　*The Times*, 12 January 1987.

19.　Greenwood, 'Defence', p. 117.

20.　This is a phrase used in the *Third Report*, Vol. 1, p. xli.

21.　See C. J. Hitch and R. N. McKean, *The Economics of Defence in the Nuclear Age* (London: Oxford University Press, 1960).

22.　*The Third Report*, Vol. 1, p. xli.

23.　*The Times*, 6 October 1987.

24.　*The Times*, 5 October 1987.

25.　L. Brittan, *Defence and Arms Control in a Changing Era*. (London: Policy Studies Institute, 1988). See also memorandum on 'The Surface Fleet and Defence Priorities within the Context of a Regular Review Process', submitted by Dr John Baylis to the House of Commons Defence Committee in March 1988: *The Future Size and Role of the Royal Navy's Surface Fleet*, Sixth Report of the Defence Committee. HC 309, Session 1987–88 (London: HMSO, 1988).

26.　The system of Long-Term Costings could be made to fit in with regular five-year overviews with little difficulty.

27.　Confidential interview, November 1987.

28.　W. R. Schilling, 'The Politics of National Defense: Fiscal 1950', in W. R. Schilling, P. Y. Hammond and G. H. Snyder, *Strategy, Politics and Defense Budgets* (New York: Columbia University Press, 1962), p. 14.

29.　Ibid, p. 15.

30.　See *The Sunday Telegraph*, 14 February 1988. It was reported that 'the Chiefs of Staff, faced with the cancellation of critical defence programmes, had shelved their traditional rivalries to present a united front to ministers'. As a result defence ministers accepted the need to present to the Cabinet a 'shock list' of politically sensitive defence programmes which 'would face cancellation unless a budget increase was forthcoming' in late 1988 or early 1989. It was clear that Mr Younger's decision to recognise the funding crisis was an effective U-turn by the Defence Minister, who had previously insisted that all programmes and commitments could be sustained within the existing budget.

31.　See *The Independent*, 24 June 1988.

32.　See *The Independent*, 2 November 1988.

10　Striking the Right Balance for the Future

1.　See *The Sunday Times*, 21 February 1988.

2.　Confidential interviews, November 1987.

3.　See K. Booth and J. Baylis, *Britain, Nato and Nuclear Weapons: Alternative Defence Versus Alliance Reform* (London: Macmillan, 1988).

4.　The author has attempted to consider these arguments in 'Britain and the Bomb', in G. Segal, E. Moreton, L. Freedman and J. Baylis, *Nuclear War and Nuclear Peace*, 2nd edn (London: Macmillan, 1988).

5.　Whether President Reagan agreed at Reykjavik to strive for the elimination of all nuclear weapons or just all ballistic missiles is not altogether clear. He did, however, make an unambiguous call for the elimination of all nuclear weapons before the Moscow summit in May 1988.

6. *Discriminate Deterrence: Report of the Commission on Integrated Long-term Strategy*, January 1988.
7. The Second Report from the Expenditure Committee, Session 1975–6 (London: HMSO, 1976).
8. K. Waltz, *The Spread of Nuclear Weapons: More May Be Better*, Adelphi Paper No. 171 (London: IISS, 1981).
9. The case for Trident is based on certain very strict criteria, like the ability to hit Moscow. If the criteria are loosened then a case can be made for a cheaper (but nevertheless effective) system.
10. See *The Guardian*, 2 October 1987.
11. *The Way Forward* (London: HMSO, 1981) Cmnd 8288, p. 5.
12. The Nimrod early warning system was subsequently cancelled and the government opted for the American AWACS system.
13. For an excellent study of the background to the adoption of flexible response see J. E. Stromseth, *The Origins of Flexible Response* (London: Macmillan, 1988).
14. *The Independent*, 19 September 1987.
15. Quoted in M. Herson and D. Smith, *We Shall Not Be MIRVed* (London: Campaign for Nuclear Disarmament, 1975).
16. Earl Mountbatten, 'The Final Abyss', in *Apocalypse Now?* (London: Spokesman, 1980).
17. *Diminishing the Nuclear Threat* (London: BAC report, 1984).
18. For a more detailed examination of this concept see the author's 'Nato Strategy: The Case for a New Strategic Concept', *International Affairs*, vol. 64, no. 1, 1987–8, and Booth and Baylis, *Britain, Nato and Nuclear Weapons*. (London: Macmillan 1989).
19. For a discussion of the WEU Report see *The Guardian*, 25 November 1987. See also J. Dean, *Watershed in Europe: Dismantling the East–West Military Confrontation* (Lexington, Mass: D. C. Heath, 1987).
20. *The Times*, 6 October 1987.
21. *The Independent*, 26 November 1987.
22. In the light of Soviet debates about defensive deterrence more consideration could be given in the West to some of the ideas of non-provocative defence. See J. Snyder, 'Limiting Offensive Conventional Forces: Soviet Proposals and Western Options', *International Security*, vol. 12, no. 4, 1988.
23. *The Times*, 6 October 1987.
24. See *The Independent*, 4 February 1988.
25. Ibid.
26. See Snyder, 'Limiting Offensive Conventional Forces'.
27. John Nott has argued that 'increasingly we fancy ourselves as a world policeman', *The Times*, 6 October 1987.
28. Ibid.

Index